A NATURALIST'S GUIDE TO THE
BIRDS
AUSTRALIA

Dean Ingwersen

JOHN BEAUFOY PUBLISHING

Photo Credits
All photos by **Dean Ingwersen** except as detailed below.
Front cover: *main image* Singing Honeyeater; *bottom left* Double-eyed Parrot; *bottom centre* Australian Pied
Oystercatcher; *bottom right* Splendid Fairy-wren.
Back cover: Mallee Emu-wren.
Title page: Western Spinebill.
Contents page: Dollarbird.
Main descriptions: photos are denoted by a page number followed by t (top), b (bottom), l (left) or r (right).
Ash Herrod 100tl. **Geoff Jones** 19b, 22t, 23t, 23b, 25t, 31t, 32t, 36t, 40b, 55t, 61b, 62t, 62b, 79t, 79b, 84b,
85b, 91b, 92t, 102t, 107t, 122b, 128bl, 132t, 134b, 141b. **David Parker** 14t, 60b. **Mick Roderick** 27t, 29b,
30b. **Andrew Silcocks** 95b. **Chris Tzaros** 13b, 17b, 26t, 31b, 35b, 46t, 47t, 65b, 66t, 75t, 76t, 83b, 88b, 89b,
92bl, 93t, 95t, 97b, 105tr, 105b, 116t, 120bl, 122t, 125b, 126t, 131b, 132b, 136t, 136b, 146t, 148b, 150t. **Jan
Wegener** 18b, 72b, 76bl, 77t, 82b, 98t, 105tl, 109b, 113b, 115b, 127tl, 143t, 147tl.

ISBN 978-1-912081-24-0

Edited by Krystyna Mayer
Designed by Gulmohur Press, New Delhi

Printed and bound in Malaysia by Times Offset (M) Sdn. Bhd.

·CONTENTS·

Acknowledgements

I would like to thank John Beaufoy for the opportunity to publish this book, Rosemary Wilkinson for her patience and professionalism in helping me produce the finished product and Krystyna Mayer for editing the text. I am indebted to David Parker and Mick Roderick for dramatically improving drafts of the manuscript and helping with species selection, never an easy task with so many cool birds to choose from. I am also indebted to the photographers who generously provided their images: Chris Tzaros, Geoff Jones, Jan Wegener, Mick Roderick, Ash Herrod, David Parker and Andrew Silcocks. I have had the pleasure of working in the field with nearly all of you, and never stop learning from you or being inspired by your work. Additionally, this book would not have been possible without the support of my wife Hollie, and our sons Liam and Ethan, in whom I hope to inspire a love of the natural world. I am forever grateful for your patience and understanding as I chase off after that next bird to see or photograph.

Introduction

Australia is an ancient, weathered landscape of amazing beauty. Once home to polar dinosaurs and dramatic megafauna that included giant flightless birds (the dromornithids), the environment has been steadily drying since separation from Gondwanaland, and the isolation has proved a fertile ground for high levels of endemism. This book provides an introduction to 280 bird species found in Australia, which have been chosen to show representative examples from all of the bird families in the country, as well as to demonstrate the diversity in some of the endemic families. The focus has also been on birds found within striking distance of the capital cities of the east coast, and those with widespread distributions that reflect the diversity of the landscapes of Australia. Further to this, information is provided on the climate and biogeography that has helped shape the avifauna now present, as well as tips on some of the best spots in Australia to see birds.

Climate

The climate of Australia is defined by the fact that the continent is broad, flat, largely geologically stable, and in a zone of many atmospheric highs and low rainfall. The rainfall across much of Australia is so low that the vast majority of the continent is classified as being arid or semi-arid, with about 50 per cent of the land mass estimated to receive less than 300mm of rainfall per annum. However, tied to this is the fact that the rainfall that does arrive in these areas is infrequent and irregular, and at times those annual rainfall figures can be obtained in one storm event.

Around the coast the climate is somewhat more predictable, with the northern regions of Western Australia, the Northern Territory and Queensland having a tropical climate with large rainfall over the hot and humid austral summer (the 'wet season'). This is followed by a mild austral autumn–winter period (the 'dry season'), which is characterized by lower rainfall and drying vegetation. However, this region is also subject to cyclones every year, which can flatten vegetation and make survival of many birds difficult – at these times birds like cassowaries can be found wandering through gardens in search of food. Along the eastern seaboard the climate on the coastal side of the Great Dividing Range is wetter than inland of the ranges, with the central regions of the east coast being subtropical and mild. In the south-west and south-east of the mainland, and in Tasmania, the climate follows more traditional and predictable patterns, with four seasons recognized (spring, summer, winter and autumn). Rainfall typically arrives in cold fronts from the southern ocean in winter and spring, with summer heat waves determined by hot winds blown in from arid northern regions.

Droughts wrought by El Niño-Southern Oscillation events, where changes to ocean currents and atmospheric conditions across the Indo-Pacific basin impact temperature and rainfall in Australia and elsewhere, have severe impacts on the country. Rain-producing tropical systems tend to fall over the ocean rather than land, warming the seas, and causing dry 'wet seasons' in the north of Australia and hotter summers in the south. This leads to increased frequency of water shortages and bush fires, both of which impact wildlife populations. During these events birds inhabiting temperate areas can suffer to the point where they fail to breed or even moult, while arid-zone species tend to get pushed to coastal refuges.

VEGETATION

Australia has been geologically stable for a very long time. As a result the soils across the continent are very old and thin, and many nutrients have been leached out of the system so that there is very low fertility in most regions. The vegetation has therefore had to adapt to obtain nutrients, and nitrogen-fixing leguminous plants have become key to ecological communities. Given the vastness of the land mass, at over 7.7 millon sq km, and the wide climatic envelope of the country, there has been considerable opportunity for evolution to shape the vegetation.

It is often considered that Australia is the land of eucalypts, and at least in temperate regions this is supported by the fact that *Eucalyptus* is the dominant tree genus. 'Gum' trees of one form or another are found across many areas, from Snow Gums (*E. pauciflora*) of the Alpine Regions, to the Karri (*E. diversicolour*) forests of south-west Western Australia and the Northern White Gum (*E. brevifolia*) of northern Australia. One of the eucalypts is among the tallest flowering plants in the world – the Mountain Ash (*E. regnans*) of the wet forests of the south-east mainland and Tasmania.

While eucalypts are often recalled when discussing Australian vegetation, the wattles (*Acacia* spp.) are more numerous and widespread, particularly across arid and semi-arid zones. The most widespread *Acacia* species include Brigalow

Spotted Gum forest, southern NSW

(*A. harpophylla*) and Mulga (*A. anuera*), both of which can form vast woodlands and associated arid scrubs, intermixed with a number of key arid plants such as *Eremophila* that support nomadic birds like Black and Pied Honeyeaters. Acacias also occur in temperate to arid regions, and several species have evolved to thrive in rainforest environments.

Both of these plant families are known as sclerophyllous, where the thick-skinned leaves are adapted to retain moisture in harsh, dry weather. In some species of acacia the leaves are even rolled to keep the stomata from being too exposed to drying conditions. It is also worth noting that the vegetation of Australia is evergreen. Due to a lack of harsh winters like those in the northernmost regions of the northern hemisphere, Australian vegetation has not suffered the evolutionary pressure to deal with extreme cold. Because of this, most habitats in Australia look superficially identical all year round.

The northern regions of Australia share flora with New Guinea and Southeast Asia, but the flora is generally typical of tropical regions across the globe – large trees with broad leaves to catch available sunlight, and various thickets of vines, shrubs, ferns and epiphytes in the understorey. Food availability for fauna is more dependent on fruiting trees in these regions, although savannah still retains flowering eucalypts that produce nectar flows. Across the arid interior there is also a suite of species adapted to irregular rainfall, such as the bluebushes (*Chenopodium* spp.) and saltbush (*Atriplex* spp.), which form integral components of vegetation communities.

As a general rule, particularly outside the tropical regions of Australia, the Myrtaceae plant family dominates the Australian landscape. Along with the aforementioned *Eucalyptus*, genera such as *Callistemon*, *Melaleuca* and *Leptospermum* are widespread and contribute a large proportion of the country's flora. Many members of the Protaceae, such as *Banksia* and *Grevillea*, are also found widely across Australia. Common to most of these

Savannah woodland, northern Australia

species is the delivery of nectar flows within their inflorescences, which is an evolutionary strategy to attract birds and mammals. This in turn has helped to shape the fauna of Australia, with a radiation in animals like honeyeaters and possums to allow access to the floral resources.

In relation to terms used for the descriptions of habitat used by birds in the species accounts, terminology follows the widely accepted classification first developed by Specht (1970), where canopy cover determines the name given. Treed habitat with a percentage canopy cover of more than 30 per cent is known as forest (over 70 per cent is closed forest); under 30 per cent is woodland; 'treeless' shrubby habitat with a percentage canopy cover of more than 30 per cent is known as scrub (or heath if under 2m tall), while under 30 per cent is known as shrubland.

BIOGEOGRAPHY

As the continents of the world started to drift apart during the Cretaceous period, Australia was linked to a number of continents – embracing South America, Africa, Madagascar, Antarctica, India, New Zealand and New Guinea – in the giant land mass known as Gondwanaland. At that time the land mass was much further south than it is today, and the climate was much wetter. Over the subsequent 135 million years the continents of Gondwana have drifted north, and for Australia and other countries this has resulted in a significant drying of the landscape.

Much of Australia's biota has its roots in that time, and as a result many families of plant and animal are also found across the aforementioned countries. For example, the flightless family of birds known as ratites has representatives in South America (rheas), Madagascar (extinct elephant birds), Africa (ostriches), Australia and New Guinea (cassowaries and Emus), and New Zealand (extinct moas). Other bird families, such as parrots, share similar origins and distributions.

When Europeans first arrived in Australia the species they encountered shared similar appearance and ecology to common birds of England and other parts of Western Europe, and names such as robin, thrush and warbler were bestowed upon the newly discovered species. It was also thought for a long time that the northern hemisphere was the basal point from which all songbirds evolved. However, molecular studies in recent decades have flipped this notion on its head, with songbirds of the northern hemisphere now accepted by most to have evolved from Australian, or at least Gondwanan, ancestors. Subsequently the songbird families in Australia are all very old, or are mostly descended from old lineages.

Within Australia itself, the gradual northwards drift and drying has seen once-fertile forests and swamplands replaced by vegetation adapted to lower fertility driving speciation. The role of bush fires as part of this cycle is also important, with many plants evolving strategies to survive fire events. As a result of these factors the fauna of the country needed to adapt, resulting in high levels of endemism across Australia. Songbirds in particular, like grasswrens, fairy-wrens and honeyeaters, are all a product of the shifting geographic envelope of Australia.

The drying of the continent has driven evolution further within already endemic families. As an example, the Nullarbor Plain, a vast, treeless arid region that straddles the middle of the southern part of the mainland now separates the temperate eucalypt forest and woodland of south-west Western Australia from structurally similar vegetation in south-east Australia. With a lack of a viable vegetation connection between those two regions, the flora and fauna of a once-contiguous region have been able to evolve independently – in the case of south-west Western Australia to the point of being recognized internationally as one of the major biodiversity hotspots. There are congeners on both sides of the Nullarbor Plain – Rufous Treecreepers in Western Australia and Brown Treecreepers in the east; Western Yellow Robins and Eastern Yellow Robins; and white-tailed black-cockatoos of Western Australia and Yellow-tailed Black-Cockatoos of the south-east. Again, there are high levels of endemism due to long isolation, with species such as Red-capped Parrot, Western Thornbill and Noisy Scrub-bird being found nowhere else on the planet.

Similar mechanisms have operated around the country, with isolated regions across northern Australia all operating to yield high levels of endemism. In these areas, such as the Kimberley Ranges of Western Australia and the Cape York region of north Queensland, habitat has become more isolated by climate shifts. As a result the rainforest of north-east Queensland, for example, is now recognized as another global biodiversity hotspot with a suite of endemic birds found there.

Gibber plains of central Australia

Where to Find Birds in Australia

Australia is a relatively easy place to get around and there are multiple highly skilled tour operators who can provide excellent value for money and service when searching for birds, although for many species a guide is not necessary. A quick, although definitely not exhaustive summary of the best birding locations in each state is given below.

WESTERN AUSTRALIA (WA)

The south-west corner of the state is recognized globally for its high level of endemism. A tour of locations such as the hills outside Perth, Dryandra Woodland, Albany and coastal regions around Margaret River will yield nearly all endemic species of the area (such as Western Bristlebird and Rufous Treecreeper). The Broome region offers excellent birding, but is famed for the migratory shorebirds visiting Roebuck Bay. To the north-east of here stretches the Kimberley's, where the sandstone-escarpment species like

Roebuck Bay, Western Australia

Sandstone Shrike-thrush and White-quilled Rock-Pigeon can be found, and further north the Mitchell Plateau is home to the Black Grasswren.

NORTHERN TERRITORY (NT)

The highlight of birding in this area is Kakadu National Park and Arnhem Land. Kakadu is perhaps best known for its vast wetlands and waterbirds, but the savannah woodland here and in the broader Arnhem escarpments is home to some great species like White-lined

Honeyeater and White-throated Grasswren. It is important to note, however, that Darwin itself is home to an amazing array of species. The centre of the state, based around Alice Springs, is also good for arid-zone species birding. In recent years an irruption of the extremely rare and nomadic Princess Parrot occurred on indigenous land close to Alice Springs.

QUEENSLAND (QLD)

The tropical north-east is high on the list of places to visit, and is home to a suite of endemic species like Golden Bowerbird, Mountain Thornbill and Bridled Honeyeater. Cairns is well situated as a place to start from; further north offers Cape York and to the west the drier Atherton Tableland provides great woodland birding. Tours to the Great Barrier Reef islands in north Queensland allow access to warm-water seabirds like Sooty and Bridled Terns. Inland Queensland provides opportunities to see species of arid Australia, and the recent rediscovery of the Night Parrot in the far south-west has generated much excitement. Closer to Brisbane a suite of endemics – such as Spotted Bowerbird, Hall's Babbler and Bourke's Parrot – can be found around places like Bowra Sanctuary.

NEW SOUTH WALES (NSW) AND
AUSTRALIAN CAPITAL TERRITORY (ACT)

Subtropical north-east New South Wales is home to Albert's Lyrebird, Regent Bowerbird and Green Catbird, with sites like Dorrigo National Park worth visiting. The sandstone escarpments of the Wollemi National Park and surrounds are the place for the endemic Rockwarbler, and the temperate woodland on the inland slopes and plains of the Great Dividing Range is the best place for Regent Honeyeater and Superb Parrot. The riverine plains are home to the Plains-wanderer, one of the most taxonomically unique species in the world, and further west arid-zone species like Redthroat, Grey Grasswren and Gibberbird can be found.

VICTORIA (VIC.)

The East Gippsland region is well known for forest owls such as Sooty, Masked and Powerful Owls. Forests to the east of Melbourne allow easy access to wet-forest birds such as Superb Lyrebird and Pilotbird, while the north and north-east offer great woodland birding for species like Turquoise Parrot, Gilbert's Whistler and Black-chinned Honeyeater. The mallee areas of the north-west are home to Mallee Emu-wren, Striated Grasswren and Red-lored Whistler. A mecca for birders in Australia

Spinifex grass and mallee eucalypts, western Victoria

is the Western Treatment Plant, where most of Melbourne's waste water is treated. The bird list for the site betters Kakadu in the Northern Territory, and along with resident and regular migratory shorebirds, it is well known for a vast array of sightings of vagrant shorebirds such as Hudsonian Godwit and Stilt Sandpiper.

TASMANIA (TAS.)

In summer Swift Parrots can be found breeding in flowering eucalypt forest of south-east Tasmania, with Bruny Island one of the most regular sites. In fact it is possible to see all 12 of Tasmania's endemics on Bruny Island, or within a day from the state's capital, Hobart. The coast contains good sites for resident shorebirds, and the remote south-west is the breeding range for the Critically Endangered Orange-bellied Parrot.

SOUTH AUSTRALIA (SA)

The Coorong wetlands of south-east South Australia are still home to thousands of migratory waders each summer, even though there has been a substantial decrease in the numbers of waders using the site due to mismanagement of the Murray-Darling river systems, as a result of which the Coorong is now in poor condition. The mallee region, particularly north of Waikerie, is probably the best spot for seeing the Critically Endangered Black-eared Miner and a host of other mallee species, and in recent years it has yielded regular sightings of Scarlet-chested Parrot (usually found in the west of the state in the Great Victoria Desert). The Lake Eyre basin is a highlight following flooding rain in south-west Queensland, and it fills with breeding Pelicans, Banded Stilts and other shorebirds.

PELAGIC BIRDING

There are a number of high-quality pelagic birding trips run around the coast, with locations including Brisbane (Qld), Port Stephens, Sydney and Eden (NSW), Portland and Port Fairy (Vic.), Eaglehawk Neck (Tas.), Port Macdonnell (SA), and Esperance and Albany (WA). These trips are run by very experienced operators and highly skilled organizers, and offer observers the chance to see a range of pelagic species of subtropical and southern oceans up close.

SAFETY NOTE

It should be noted that Australia is large and most of the population resides within close proximity to the coast. Once outside towns and cities in arid and semi-arid regions, people need to take precautions and consider that they are travelling and birding in remote locations. Necessary precautions with vehicles, communications and emergency supplies need to be considered – there are still people who unfortunately perish each year in the Outback because they were not prepared for what the landscape would offer.

Where to Submit Records

BirdLife Australia coordinates a long-running Atlas of Australian Birds project. Running since 1998, there are now well over 12 million species records in the database, which is used by policy makers, land managers, researchers and community members. It is now being used to develop the first national bird indices, which will inform future management of threatened bird species and sites. Preferred methods include 20-minute x 2ha surveys, but incidental and area search results are also encouraged. Birders can register to submit records via email: atlas@birdlife.org.au. An increasing number of birders are submitting records to eBird, an online repository coordinated by the Cornell Lab of Ornithology: www.ebird.org.

Topography of a Bird

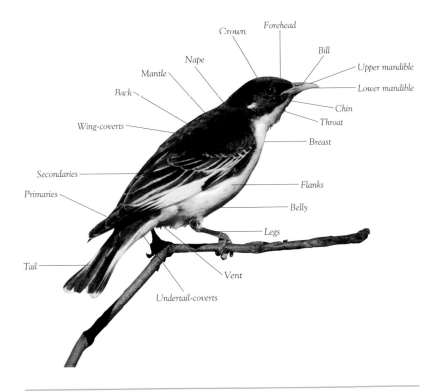

Southern Cassowary
▪ *Casuarius casuarius* 1.5–1.8m

DESCRIPTION The second largest native bird in Australia, and unmistakable. Adults have black plumage that hangs shaggily. Bare skin of head is pale blue colour that continues down sides of neck, deepening to dark blue, to purple. Vibrant pink-red wattles hang under throat; bony helmet on top of head. Flightless. **DISTRIBUTION** Restricted to coastal north Qld, from rainforests of Cape York south to near Townsville. **HABITS AND HABITAT** Found in thick rainforest, especially areas around streams, creeks and clearings. Also occurs around rainforest edges, where it may wander into gardens and yards, and come into conflict with pets like dogs. Generally solitary. Male incubates eggs alone and raises young for up to nine months; female may mate with multiple males.

Emu ▪ *Dromaius novaehollandiae* 1.5–2m

DESCRIPTION Australia's largest native bird, and flightless. Loose, shaggy, grey-brown plumage comprising double-shafted feathers on body. Dark feathers on head, with bare blue skin down each side of head and neck. Small vestigial wings can be seen on sides of body. **DISTRIBUTION** Right across mainland Australia, although generally absent in true deserts. Seems to have seasonal movements in south-west WA. Tasmanian subspecies extinct since European settlement. **HABITS AND HABITAT** Well adapted to semi-arid and rangeland areas, and becoming less common around towns and large cities. Somewhat nomadic, at times moving great distances in response to rain and better conditions. Fairly shy and retreats if startled, but at times quite curious. Male alone broods and raises chicks for as long as 18 months.

Australian Brush-turkey ■ *Alectura lathami* 60–70cm

DESCRIPTION Distinctive ground-dwelling bird. Head largely bare bright red skin that continues down neck to large yellow wattle at base. Back and wings dull black. Long tail flat and shaped like a fan. **DISTRIBUTION** Along coastal eastern Australia and offshore islands, patchily distributed from tropics of Cape York as far south as Sydney region. Occurs inland as far as Pilliga region of NSW and Charters Towers in Qld. **HABITS AND HABITAT** Favours tropical subtropical and temperate rainforests and scrubland. One of several mound-nesting species in Australia that lay their eggs in mounds of decomposing vegetation for incubation. Both sexes tend to nest, adding or removing vegetation to maintain correct temperature. Also found in parks and gardens in some areas. Young independent from the moment of hatching.

Malleefowl

■ *Leipoa ocellata* 55–60cm

DESCRIPTION Beautiful plumage that provides excellent camouflage. Upper body has scalloped and barred patterns of brown, white, black and chestnut plumage; grey neck broken by line of black feathers on throat. Washes of orange-chestnut in face and underparts. **DISTRIBUTION** Occurs in inland scrub and mallee areas from central south-western slopes of NSW, north-west Vic. and southern SA. Also found across wheat belt of south-west WA. Formerly more widespread. **HABITS AND HABITAT** Very shy and wary, and hard to approach closely. Mound-nesting species; its mounds can be up to 5m across and more than 1m tall. This is where most sightings usually occur. Female lays 15–30 eggs in nest mound made and maintained by male, which scratches vegetation on or off to keep the temperature stable.

Stubble Quail ■ *Coturnix pectoralis* 16–20cm

DESCRIPTION Upperparts rich grey-brown with white streaks within plumage, as well as brown, buff and grey striations and long white eyebrow. Male has rich orange-buff throat

and black patch on breast; female has white throat. Both sexes cream below with black. **DISTRIBUTION** Found from southern Cape York all the way around south-east of Australia to central SA, including Tas.; also south-west WA. Irruptive in good seasons and more nomadic through central and northern Australia, where it is seen less often. **HABITS AND HABITAT** Calls of birds usually first sign of their presence, and they are found in singles, pairs or small coveys. Like most quail will explode off ground when disturbed, flying low and fast before skidding into vegetation. Occurs in a variety of habitats, including grain crops and stubble, grassland and rangeland. Also found in areas of saltbush, bluebush and spinifex.

Brown Quail ■ *Coturnix ypsilophora* 17–22cm

DESCRIPTION Australia's largest native quail. Rich chestnut-brown plumage overall, with fine faint white streaks on upperparts. Face plain with large dark cheek-spot and red

eyes; Tasmanian race has yellow eyes. Flanks and underparts plain chestnut with dark brown chevron patterns. Darker phase exists where upperparts contain black barring. **DISTRIBUTION** Distributed widely but patchily around coast of mainland, except central WA coast and southern SA, and in Tas. **HABITS AND HABITAT** Reflective of its distribution, utilizes a wide array of habitats – long grass near wetlands and watercourses, crop stubble, green crops, coastal heathland, tropical grassland and savannah, and bracken. Often found in singles to small coveys; tends to be most active at dawn and dusk, and often seen on roads and tracks. Flushed birds burst out of vegetation in all directions, fly quickly and land rapidly in cover.

Magpie Goose ■ *Anseranas semipalmata* 75–95cm

DESCRIPTION Distinctive appearance. Reddish facial skin and bill before knobbed black head. Neck, upperwings and tail black; underparts and most of underwing white, giving noticeable pied plumage. Long, orange-yellow legs. **DISTRIBUTION** Formerly widespread across northern and eastern Australia; now restricted to coastal regions of northern NSW, Qld, NT and northern WA. Reintroduced into parts of Vic. and SA, self-establishing in several regions of NSW riverina. **HABITS AND HABITAT** Favoured habitat is large wetlands, particularly those that are seasonal or ephemeral in nature. Also found on floodplains, and rush and sedge-dominated swamps and dams. Occasionally occurs on rice crops near key wetlands. Can be seen in flocks numbering in their hundreds, particularly when breeding in large colonies. Persecuted for food following European settlement, hence the need to reintroduce birds in southern part of range.

Plumed Whistling-Duck ■ *Dendrocygna eytoni* 40–60cm

DESCRIPTION Crown and rear neck pale brown, and throat and neck cream. Blotched pink bill. Upperparts greyish-brown, and breast rich rufous with black stripes. White belly and long pink legs give tall, 'goose-like' appearance. Distinctive buff-coloured flank plumes, fringed with black. **DISTRIBUTION** Across northern and eastern Australia in broad arc from northern WA right around to northern Vic. Tends to disperse or migrate seasonally to some areas. **HABITS AND HABITAT** Favoured habitat includes margins of wetlands and waterholes (both permanent and ephemeral), farm dams, floodplains and irrigated land, where it is often found roosting in large flocks during the day, while at night it forages on grasses and seeds. Also found on sewage farms and in rice fields. In flight gives distinctive whistling call.

Freckled Duck ■ *Stictonetta naevosa* 50–59cm

DESCRIPTION Plumage blackish-brown overall, with buff to white 'freckles' or fine barring. Small crest of feathers on crown gives head a peaked shape, while bill is long and flattened towards tip. Male develops crimson flush on base of bill in breeding condition.

DISTRIBUTION Core distribution is in inland south-east Australia, and also in south-west WA. Disperses to more coastal regions of eastern and southern Australia in response to changes in conditions; similar in north-west WA. **HABITS AND HABITAT** Generally found in flocks of several individuals up to hundreds of birds. Often roosts on rocks, posts and emergent timber in waterbodies. Forages in open water, where it upends to filter feed. Favours large, well-vegetated swamps and wetlands, although at times can be found in good numbers on farm dams and lakes in towns and cities.

Cape Barren Goose ■ *Cereopsis novaehollandiae* 75–90cm

DESCRIPTION Large grey goose with heavy-bodied appearance. White crown, and short black bill with large, distinctive lime-green cere. Wing feathers have dark spots.

Reddish legs and square black tail. **DISTRIBUTION** From north-east Tas. through islands of Bass Strait to central Vic., then across coastal SA and offshore islands. Second race exists on islands off southern coast of WA, with some birds present on mainland in that area. **HABITS AND HABITAT** Commonly found on offshore islands, among tussock grassland and areas of scrub. Also occurs on improved pasture, at times congregating into large flocks. Can be seen in and around large sewage farms and dams on mainland. Some post-breeding dispersal from offshore islands to mainland areas is known.

Black Swan ▪ *Cygnus atratus* 1.1–1.4m

DESCRIPTION Notable as the world's only mostly black swan, this species has a distinctive appearance. All feathers are black over body, with the only white feathers being in outer wing, which is easily seen in flight and often seen when resting. At rest wing-coverts can appear ruffled or wavy. Deep red bill with white band at tip of upper mandible. **DISTRIBUTION** Distributed widely across Australia, including Tas. Less common in central and northern parts, although there is some evidence of range expansion in some regions. **HABITS AND HABITAT** Found from singles to large flocks numbering thousands, a scene encountered often by early explorers. Goes through flightless stage during post-breeding moult, when large numbers can be seen on secure waters like large freshwater lakes, its preferred habitat. Also uses brackish and salt water, and commonly encountered in man-made lakes in towns.

Australian Shelduck ▪ *Tadorna tadornoides* 55–75cm

DESCRIPTION Distinctive dark-plumaged bird with white, buff and green panels in upperwing (which can be seen at rest), and white underwing in flight. Both sexes have white collar around neck, above buff breast in male and rich chestnut breast in female. In female, white eye-ring and plumage around base of bill. **DISTRIBUTION** Seen in greatest densities in western Vic. and south-west WA, but also found in bottom part of Murray-Darling Basin, coastal NSW and Tas. **HABITS AND HABITAT** Grazing species often seen around large, shallow wetlands where it can upend in the water in search of food. Often loafs in large rafts on open water. Favours freshwater habitats, but can be found in brackish, saline and tidal areas, even occurring on open sea at times. May disperse over large distances after breeding.

Australian Wood Duck ■ *Chenonetta jubata* 45–50cm

DESCRIPTION Pretty grey duck with two black stripes down back and an ornately spotted, brown-and-white breast (which extends to belly in female). Female has buff-coloured head with pale stripe above and below eye, while male has dark brown head and small black mane on nape. **DISTRIBUTION** Found widely across eastern half of Australia, including Tas., and in south-west WA. **HABITS AND HABITAT** Grazing duck, most often seen in grassland and irrigated crops; appears to have benefited from agricultural expansion. Also utilizes open woodland, farm dams and wetlands, foraging in shallow water, crops, coastal bays, and town parks and ornamental gardens, where it is easily encountered.

LEFT Male; RIGHT Female

Pink-eared Duck ■ *Malacorhynchus membranaceus* 37–45cm

DESCRIPTION Stunning bird with broad, brown-and-white stripes down sides, giving rise to alternative name of Zebra Duck. Wings and upper tail plain brown, while undertail has buff wash.

Head distinctive, with large brown patch around eye sitting inside whitish face. Small patch of pink feathers just behind eye, and large, squarish bill. **DISTRIBUTION** Widely spread across south-eastern and south-western Australia, mostly in inland areas. Highly mobile and nomadic, moving and breeding in response to rainfall. **HABITS AND HABITAT** Usually seen in small to large flocks, floating across open water with bill submerged as it filter feeds. Found in wide variety of habitats, from ephemeral inland floodplains and claypans, to sewage farms and commercial saltfields. Favours fresh water, though can tolerate very saline waters.

Chestnut Teal ■ *Anas castanea* 35–50cm

DESCRIPTION Male in breeding condition beautifully plumaged, with iridescent green head, and rich chestnut throat, breast and underparts. Feathers of underparts also have dark brown centres, with rear of bird terminating in white stripe before dark tail; brown wings. Female and eclipse-plumage male mottled buffy-brown with pale throat. **DISTRIBUTION** South-east Australia from bottom of Cape York in Qld around to Eyre Peninsula in SA, including Tas. and offshore islands. Also south-west WA. **HABITS AND HABITAT** Found in wide variety of habitats, although prefers saltwater or brackish conditions. Utilizes coastal estuaries, lakes, marshes and mudflats. Also seen on freshwater lakes and wetlands both coastal and inland. Dabbles and upends in water, and often found in flocks roosting in trees around wetlands in association with **Grey Teal** *A. gracilis*.

Red-tailed Tropicbird ■ *Phaethon rubricauda* 85–95cm

DESCRIPTION Stunning white seabird with white (sometimes pinkish) plumage broken only by black shafts to primary feathers in wing and tertials, and black line through eye on head. Short, strong red-orange bill and long red tail streamers combine to give distinctive appearance. Tail hard to see from a distance, giving bird a stubby-tailed look. **DISTRIBUTION** Spread around offshore islands of Australia, from tip of south-west WA around top end, to central NSW on east coast. Lord Howe Island is said to have the largest breeding concentration in the world. Vagrants can get pushed into southern waters around Vic. and SA after tropical storms at times. Rarely seen on or near mainland. **HABITS AND HABITAT** Found on tropical and subtropical seas, islands and coasts. Nests under bushes or cliffs, where it makes a scrape in the ground. Usually solitary at sea, soaring and hovering high before diving into water to catch fish fully submerged.

Australasian Grebe ■ *Tachybaptus novaehollandiae* 23–25cm

DESCRIPTION Australia's smallest grebe, with a black head and black neck that has a chestnut stripe. Bare, oval-shaped yellow patch of skin at base of bill when breeding. Body

generally brown to buff. Yellow eyes. In non-breeding season plumage is much duller and yellow face-patch whitens considerably. **DISTRIBUTION** Widespread across mainland and Tas., but most common in east and south-west of the country. Uncommon and irregular visitor to inland and northern Australia. **HABITS AND HABITAT** Usually seen in territorial pairs, although does coalesce into larger flocks in autumn–winter, when not breeding. Favours still, shallow freshwater wetlands, including sewage farm ponds and farm dams. Dives when startled; carries young under wings even when diving for prey such as freshwater crayfish.

Hoary-headed Grebe ■ *Poliocephalus poliocephalus* 29–31cm

DESCRIPTION Small, plain-plumaged bird, set off in breeding season by head bearing numerous white 'plumes' above black feathers, which give a hairy appearance. Neck

pale, changing to darker body and brown wings (when folded). Head moults back to bicoloured brown and white in non-breeding season. **DISTRIBUTION** Similar distribution to Australasian Grebe (see above), being widespread across mainland and Tas., but most common, and at times abundant, in eastern and south-western regions. Scarce and seasonal in northern and central Australia. **HABITS AND HABITAT** When alarmed tends to fly away across waterbody, in contrast to Australasian Grebe, which dives. Feeds predominantly by diving and pursuing prey such as freshwater crayfish, dragonfly larvae, bugs and other invertebrates. Occurs more often on brackish and saline wetlands than Australasian Grebe. Seen in pairs or small groups, but also gregarious at times.

Brown Cuckoo-Dove
■ *Macropygia amboinensis* 39–45m

DESCRIPTION Large, long-tailed and graceful-looking dove. Upperparts rich coppery-brown; iridescent around nape and neck in male. Underparts plainer cinnamon-brown. Female also shows some faint scalloping around sides of neck. **DISTRIBUTION** Coastal eastern Australia, including islands offshore, from central Cape York to southern NSW. Rarely inland of Great Dividing Range, it shows seasonal movements in southern part of range. **HABITS AND HABITAT** Found in both highland and lowland rainforests; call one of the most common sounds in these habitats. Not shy and at times quite gregarious. Also known to utilize Brigalow scrubs, and thickets of plants like Wild Tobacco and lantana; frequents gardens.

Common Bronzewing ■ *Phaps chalcoptera* 30–36cm

DESCRIPTION Breast pink-brown, and back brown with pale fringes to feathers. Wings contain dark primaries, setting of large patch of 'bronzed' orange-and-green feathers. Male has yellowy-cream forehead, and rich brown nape and crown; female has grey forehead. White line under eye in both sexes. **DISTRIBUTION** Widespread in suitable habitat across mainland and in Tas., although south-east and south-west Australia are where it is most common. Less common in arid central regions and Cape York. **HABITS AND HABITAT** Flighty species that is quite wary. Easily disturbed, when it will fly directly to a tree and stop, bobbing its head and watching closely. Utilizes wide variety of habitats, including woodland, dry and wet forests, mallee, farmland, heaths and coastal scrub.

Flock Bronzewing ▪ *Phaps histrionica* 28–31cm

DESCRIPTION Upperparts rich cinnamon-brown; underparts grey. Primary feathers dark with white spots. Male's head black with white face, throat and ring around ear. Female has

smutty-white face and brown forehead. **DISTRIBUTION** Range covers central arid Australia, from inland north-west NSW through south-west Qld, NT and northern WA. **HABITS AND HABITAT** Enigmatic species known to quickly appear in vast numbers around farm dams and ephemeral water in arid areas, only to vanish just as quickly. Named for the fact that it can be found in flocks numbering tens of thousands, although this is less common than it was at time of European settlement. Found across arid Australia on treeless, grassy plains, floodplains and wet claypans, spinifex and open mulga.

Crested Pigeon ▪ *Ocyphaps lophotes* 30–36cm

DESCRIPTION The only grey-brown pigeon with a slender black crest; distinctive appearance. Wings variously barred black and white, and also contains patch of iridescent

green-purple feathers not unlike those of bronzewings. Red skin around eye and pinkish legs. **DISTRIBUTION** Widely spread across mainland, although traditionally restricted to arid and semi-arid zones. As a result of agricultural and urban development, has expanded towards coast in south-east Australia; now common around Melbourne. **HABITS AND HABITAT** Commonly found on agricultural lands like farms and pastoral leases; also utilizes sports grounds, homestead gardens, watercourses and roadsides. Occurs in singles, pairs and small flocks, foraging quietly on the ground. If disturbed takes to the air in explosive burst of whirring wingbeats, and does a distinctive tail flick when it lands.

Spinifex Pigeon
▪ *Geophaps plumifera* 20–24m

DESCRIPTION Tall reddish crest on top of head is distinctive. Body reddish-orange, with black barring in wings and single black band across breast. Bright red face-skin through eye, bordered above by black and below by white and black. Grey cheek-patch. The three races have varying amounts of white on abdomen and breast. **DISTRIBUTION** Spread across northern inland Australia, from central NT and western Qld, to Kimberly and broader Pilbara regions of northern WA. **HABITS AND HABITAT** Favours spinifex grassland and rocky hills, although also known to use acacia scrubs and open, grassy woodland. Always close to permanent water. Forages quietly on the ground in pairs and small groups, and can become quite tame in places like camp grounds.

White-quilled Rock-Pigeon ▪ *Petrophassa albipennis* 26–30cm

DESCRIPTION Overall blackish-brown bird in Kimberley region of WA, becoming a lighter, redder colour moving east across its range. Dark line of feathers through eye, bordered above and below by white speckling; white speckling on upper throat. In flight distinctive broad white patch in wing, as opposed to chestnut patch of **Chestnut-quilled Rock-Pigeon** *P. rufipennis* of Arnhem Land in NT. **DISTRIBUTION** Restricted to Kimberley region of north-west WA, east of Derby. Just makes it into western NT on upper Victoria River. **HABITS AND HABITAT** Closely associated with sandstone escarpments and gorges, and habitats therein like pockets of rainforest. Its calls resonate through the escarpments, where it roosts and loafs after foraging. Found in singles, pairs and small flocks. Plumage is well-adapted camouflage in favoured habitats.

Diamond Dove ■ *Geopelia cuneata* 19–24cm

DESCRIPTION Smallest Australian dove, with overall grey-blue appearance, darker above and greyer below. Belly and undertail white. Wings ornately spotted, and bright red ring

around each eye. Female appears browner than male, with paler red eye-ring. **DISTRIBUTION** Found in driest parts of all mainland states, although absent from most arid deserts. Rarely found close to coastal southern Australia. **HABITS AND HABITAT** Usually occurs in pairs or small flocks of up to 20–30 birds, which are generally quiet and unobtrusive. Walks sedately when foraging, although can move quickly and nimbly if disturbed. Commonly heard giving mournful, repeated *coo-coo* call. Favoured habitats include arid and semi-arid grassland savannah, grassy woodland and timbered watercourses.

Wonga Pigeon ■ *Leucosarcia melanoleuca* 38–45cm

DESCRIPTION Large, plump pigeon with small head and long tail. Head, neck and upperparts uniformly grey; lower breast and belly white with large black spots. Upper breast

has diagnostic white 'V' stripe across it. Pale pink bill, pale face and pink legs. **DISTRIBUTION** Coastal eastern Australia, including parts of Great Dividing Range, from around Mackay in Qld down to west Gippsland in Vic. **HABITS AND HABITAT** Sedentary species; call often heard well before it is seen. Spends most of its time on the ground, and is often reluctant to fly. Usually found singly or in pairs. Generally occurs in rainforest and vine forest, or tall, dense eucalypt forest, although also uses open woodland. Often seen around car parks and roadsides.

Wompoo Fruit-Dove ■ *Ptilinopus magnificus* 35–35m

DESCRIPTION Australia's largest fruit-dove, with a small-headed, heavy-bodied appearance. Head pale blue-grey, grading to bright green back and wings. Each wing has a line of yellow feathers. Throat has a line of purple plumage that broadens to cover breast before terminating on belly to yellowy-orange. Underwings yellow-orange in flight. The only fruit-dove in Australia with a pink bill. **DISTRIBUTION** Found along coast, including close islands, of eastern Australia from tip of Cape York in Qld to lower Hunter Valley, NSW. **HABITS AND HABITAT** For its size, can be very hard to see among foliage cover. Singles to small parties found foraging on ripening fruits. Often located via calls or observation of falling fruits, and can at times occur quite low in foliage when feeding. Favours rainforest and monsoon forest, although also uses nearby eucalypt forest and open areas containing fruiting trees.

Tawny Frogmouth
■ *Podargus strigoides* 35–50cm

DESCRIPTION Plumage variable, but most commonly seen birds are largely grey-brown with mottled or streaked dark plumage of upperparts and breast. Wings have mottled dark and light patches, and feathers around face give impression of long eyebrows when roosting. Bright yellow eyes and broad bill. Inland and northern race rufous rather than grey-brown. **DISTRIBUTION** Most widespread frogmouth in Australia, found across whole mainland, Tas. and many islands. **HABITS AND HABITAT** Pairs and family parties roost during the day in trees, where they hold themselves still and branch-like. At dusk they become active, and actively search for prey from branches and fences, before gliding silently down to the ground to catch it. Utilizes wide array of habitats, including eucalypt woodland and forest, rainforest edges, timbered watercourses, mallee, mulga, alpine woodland, parks, golf courses and residential gardens.

Spotted Nightjar ■ *Eurostopodus argus* 27–35cm

DESCRIPTION Predominantly grey, brown and buff nightjar, with much scalloping, spotting and barring. Underparts plain rufous, and white line across throat. In flight has bold white patch in wings, differing from white spots of **White-throated Nightjar** *E. mystacalis*. Distinctive call. **DISTRIBUTION** Widespread across mainland, but absent east of Great Dividing Range in south-east. Also on coastal islands; possibly migrates in winter. **HABITS**

AND HABITAT Roosts on the ground during the day, before emerging at dusk to forage. Typically seen in last light hawking above open areas in suitable habitat, for example above dams and lakes or cleared patches of vegetation. Found mostly in open eucalypt woodland, particularly on stony, sandy ridges, although also in mallee and mulga scrubs, and spinifex areas inland.

Australian Owlet-nightjar ■ *Aegotheles cristatus* 19–25cm

DESCRIPTION Small, delicate-looking bird that appears to have a disproportionately large head and eyes. Body mottled light and dark grey, with upperparts appearing darkest.

Head dark plumaged, but contains three chestnut-buff lines through face and forehead. Long feathers around bill look like whiskers, and are used for sensory foraging. **DISTRIBUTION** Widespread across Australian mainland, Tas. and offshore islands. Common in suitable habitat. **HABITS AND HABITAT** Nocturnal species, which at times can be seen roosting in tree hollows during the day. Soft feathers adapted for silent flight, which is used to prey on insects. Gives distinctive strident, churring call, usually while perched. Habitats used include rainforest, dry and wet eucalypt forests, woodland, arid inland scrub and mallee, and timbered watercourses and wetlands.

White-throated Needletail ■ *Hirundapus caudacutus* 20cm; WS 49cm

DESCRIPTION Largest swift in Australia. Body dark with white forehead, throat, flanks and undertail. Tail appears stubby and slightly rounded, as opposed to thin and tapered **Fork-tailed Swift's** *Apus pacificus* tail in flight. In good conditions paler back and iridescent green wings can also be seen.

DISTRIBUTION Regular summer migrant to Australia after breeding season in northern hemisphere. Generally favours east coast of mainland and Tas., especially eastern Qld. **HABITS AND HABITAT** Seeing this species in flight is a true delight, particularly if the birds are flying low – an audible 'woosh' can be heard as the wings cut through the air. Most often seen in humid, unsettled and thundery conditions, where it flies on shallow, quick wingbeats or long, raking glides to catch aerial insects. Can be seen from near ground level up to several thousand metres in the air. Found in small parties up to large, loose flocks, at times in association with Fork-tailed Swifts.

White-faced Storm-Petrel ■ *Pelagodroma marina* 18–21cm

DESCRIPTION This species has the distinction of being the only storm-petrel with a white forehead, face and underparts. White of face broken by black line through eye, while head, nape and back are slaty-grey. Primary and secondary feathers, and tail, sooty-black. **DISTRIBUTION** Found in coastal waters and islands of southern Australia, from Abrolhos group in WA to Broughton Island NSW. Breeds on many of these islands in spring–summer, before dispersing through either the Indian Ocean or Pacific Ocean over winter. **HABITS AND HABITAT** Pelagic species spending extensive time at sea where the birds fly into the wind, legs trailing in the water, picking invertebrates from the surface. Also feeds while swimming. Rarely seen by day around land, leaving at dawn and returning after dusk.

Wandering Albatross ■ *Diomedea exulans* 1.1–1.35m; WS 2.5–3.5m

DESCRIPTION Plumage varies considerably with age, but adults distinguished by white back, white upperwings close to body, and white underparts, except for dark primary

feathers and thin portion of trailing edge of wing. Bill large and pink, with bulbous tip on end. **DISTRIBUTION** Occurs in oceans of southern Australia, from Fremantle region of WA around to southern Qld. Not often found over continental shelf; most frequently closer to shore after storms. **HABITS AND HABITAT** Generally found singly in pelagic waters, where it glides and 'surfs' wind currents just above the waves. At times occurs in small parties, fighting over food with other albatrosses and giant-petrels on water's surface.

Shy Albatross ■ *Thalassarche cauta* 0.9–1m; WS 2.1–2.5m

DESCRIPTION Largest black-backed albatross, of which there are three races. In all races, upperwing black from wing-tip to wing-tip, and underwing white with narrow black margin around outside; distinctive black 'thumb-print' mark right next to body on front edge of

wing. **DISTRIBUTION** Found in southern oceans, mostly around south-east corner of Australia, but does occur as far north as southern Qld and mid-west coast of WA. The only albatross to breed around mainland, for example in Bass Strait, and islands south of Tas. **HABITS AND HABITAT** Sometimes seen during land-based sea watches, particularly around coastal Vic., Tas. and NSW following big seas or storms. Skims the waves, cruising in search of its favoured food of fish or cephalopods. Commonly follows fishing boats looking for scraps.

Southern Giant-Petrel ▪ *Macronectes giganteus* 0.85–1m; WS 1.5–2.1m

DESCRIPTION Very large petrel that is the same size as some of the small albatrosses. Adult birds are one of two morphs – a dark bird that is brown on body and has a whitish

neck and head, and a white morph that is white all over except for a few black feathers. Both morphs have a large, pale pinkish tubenose bill that is tipped with green. **DISTRIBUTION** Occurs mostly in winter and autumn, and can be seen around southern coast from Fremantle, WA to Sydney region, NSW. **HABITS AND HABITAT** An impressive scavenger, foraging in pelagic waters on dead whales, dolphins, seals and other seabirds. At times will follow fishing boats looking for discarded scraps, and very occasionally birds can be found resting on water close to land.

Wedge-tailed Shearwater ▪ *Ardenna pacifica* 38–46cm; WS 97–105cm

DESCRIPTION Two colour morphs, with darker birds much more common in Australia than pale morph. Dark morph dark blackish-brown all over, with wedge-shaped tail

protruding beyond feet in flight and usually held tight and pointed. Light morph pale on underbody and underwing. **DISTRIBUTION** Common breeding visitor in summer along central coasts of western and eastern mainland. Considered rare or vagrant in northern and southern waters. **HABITS AND HABITAT** The common 'muttonbird' of warmer waters around the coast. Along with other shearwater species, can suffer from huge 'wrecks' in some years, where birds return to Australia in such poor condition that they die in the tens of thousands and wash up on beaches around Australia. Occurs from well inshore to truly pelagic waters.

Short-tailed Shearwater ■ *Ardenna tenuirostris* 40–45cm; WS 95–100cm

DESCRIPTION Another of the uniformly dark brown shearwaters, although this species has plumage that can appear reflective and give appearance under wings of being pale in

patches. Usually has pale throat, and feet protrude beyond tail in flight. **DISTRIBUTION** Spread around coast and islands of south-east Australia, including Tas. Arrives in late spring and departs to northern hemisphere in early autumn. **HABITS AND HABITAT** The common 'muttonbird' of cooler southern waters, and still harvested on islands of Bass Strait between mainland and Tas. Occurs in huge nesting colonies; breeding birds gather in numbers offshore late in the day, and wait for darkness to arrive before descending en masse to individual burrows to feed chicks. Tasmanian population alone has been estimated at more than 16 million birds.

Providence Petrel
■ *Pterodroma solandri* 40cm; WS 95–105cm

DESCRIPTION Large, heavily built bird with stout tube-nose bill. Appears mainly dark grey-brown, with grey mantle, scaly white face and white underside of primary feathers that show in flight. **DISTRIBUTION** Breeds on Lord Howe Island, and formerly bred on Norfolk Island. Most often seen off coast of Tas., eastern Vic. and NSW in winter–spring. **HABITS AND HABITAT** Pelagic species most at home on open ocean, but at times ventures in over continental shelf. Its common name reflects the bounty it provided for early European settlers and convicts, who harvested it out of existence on Norfolk Island in the late 1700s. Also suffered from impact of introduced pigs and goats. Some birds migrate across Pacific Ocean to overwinter offshore of North America.

Gould's Petrel

▪ *Pterodroma leucoptera* 30cm; WS 70cm

DESCRIPTION Small, lightly built gadfly petrel that has contrasting plumage of dark brown and grey above, and white underparts. In flight sooty-black cap on head extends down onto sides of neck; upperwings have strong dark 'M' band across them. White underwings have dark leading edge angling backwards from bend in a tapering bar. **DISTRIBUTION** Breeds on Cabbage Island off coast of Port Stephens, NSW. Seen around south-east continental shelf and beyond, mostly in summer and early autumn. **HABITS AND HABITAT** Mostly a pelagic species, but also sighted offshore at times. Usually seen singly during the day, but tends to be gregarious around nesting colonies at night. Nests in rocks and crevices, and under fallen vegetation on the ground.

Little Penguin

▪ *Eudyptula minor* 40–45cm

DESCRIPTION The smallest of all penguins, with blue-black plumage on back, upperwings and head, and white plumage on all underparts. Pink feet. Lacks any conspicuous head markings or feathers, unlike other penguins. **DISTRIBUTION** The only resident penguin in Australia, breeding around the coast and islands from north of Perth (WA), around south coast of SA and Vic., and up into NSW. Most abundant in Bass Strait and around Tas. **HABITS AND HABITAT** Well-known tourist attraction at several locations (such as Phillip Island in Vic.), where breeding colonies allow visitors close access to birds as they return to their burrows every night. Colonies still exist around capital cities like Melbourne and Sydney, although they are under increasing pressure from urban development and introduced predators such as Red Fox.

Lesser Frigatebird ▪ *Fregata ariel* 70–80cm; WS 1.75–1.95m

DESCRIPTION Typical frigatebird shape – long, deeply forked tail, long, pointed wings, and long bill with hooked tip. Male all blackish with white 'armpits' visible in flight; female has white collar, breast and 'armpits'. Male also has inflatable scarlet throat-pouch that develops during courting and breeding seasons. **DISTRIBUTION** Common around northern Australia and offshore regions such as Ashmore Island (WA), from north of Brisbane (Qld) to Kimberley coast (WA). **HABITS AND HABITAT** Like all frigatebirds, well known for

behaviour of harassing other seabirds until they regurgitate or drop their food, which the frigatebirds then catch. Also catches fish and some cephalopods from the ocean's surface itself. Breeds in colonies on oceanic islands, and sand and coral atolls.

Australasian Gannet ▪ *Morus serrator* 84–91cm; WS 1.7–2m

DESCRIPTION Generally white-plumaged bird. Trailing edge and primary feathers in wing are black, as is outer part of tail. Crown and sides of neck orangy-yellow; beak and

eye surrounded by fine line of black feathers. Small black line of feathers runs down throat just under chin. **DISTRIBUTION** Widespread around southern coasts of Australia year round, and spreads further up east and west coasts during winter. Breeds on many islands around south coast; main colony near Portland (Vic.). **HABITS AND HABITAT** To watch a flock of foraging gannets is a true sight to behold, with the birds descending at great speed, wings tucked well back behind them, to plunge beak-first into the water to catch fish below the surface. Found from singles to large flocks and colonies, and often seen cruising near beaches in search of food.

Australasian Darter ■ *Anhinga novaehollandiae* 86–94cm

DESCRIPTION Large, slender, cormorant-like bird. Male all black with rufous wash on mid-throat, and line of white plumage below eye from beak back to upper sides of neck.

In female upperparts a washed out black-grey, with white neck, breast and belly. Both sexes have long, straight, pointy, yellow-green bills. **DISTRIBUTION** Widespread across mainland, except drier central-western interior. **HABITS AND HABITAT** Distinctive kink in neck facilitates striking and impaling fish under the water – also known as Snakebird for the appearance this foraging and striking gives. Tends to favour freshwater wetlands, swamps and lakes, but can also tolerate salt water. Plumage not waterproof, so most often seen perched on stump with wings outstretched to dry.

Little Pied Cormorant ■ *Microcarbo melanoleucos* 55–65cm

DESCRIPTION Small pied cormorant, with black upperparts and white underparts. Face also white, which continues about the eye and helps distinguish it from Black-faced

Cormorant (p. 34). No coloured bare facial skin, unlike in similar but larger **Pied Cormorant** *Phalacrocorax varius*. Hooked yellow bill, and black crown with small crest on forehead. **DISTRIBUTION** Distributed widely across mainland and Tas., although like Australasian Darter (see above) absent from central-western desert regions. **HABITS AND HABITAT** Extremely adaptable, occurring in virtually all aquatic habitats from freshwater wetlands, creeks and farm dams, to coasts, inlets and estuaries. Forages by actively chasing its favoured prey of freshwater crustaceans and fish underwater, and can also follow and harass larger birds like herons, at times inflicting injuries to the larger birds while attempting to steal their food.

Black-faced Cormorant
■ *Phalacrocorax fuscescens* 61–69cm

DESCRIPTION Australia's only endemic cormorant, this bird has black upperparts and white underparts, broken only by diagnostic black line on flanks. Black cap also extends down and encircles blue eye, and bill is black. **DISTRIBUTION** Main area of occurrence around coast and islands of Tas., Vic. and South Australia, but also extends into east coast of southern WA and occasionally southern NSW. **HABITS AND HABITAT** Very much a bird of the sea, found in coastal waters and associated inlets, bays and rock stacks. Feeds on fish by plunge diving and chasing prey underwater, and has been recorded diving to depths of more than 10m. Can roost in large colonies at times, and nest is large, round, open structure lined with seaweed, driftwood and coastal debris. Generally lays 2–3 eggs.

Australian Pelican ■ *Pelecanus conspicillatus* 1.6–1.8m; WS 2.3–2.5m

DESCRIPTION Large, mostly white pelican, with only black plumage being in wings, rump and tail. Short legs. Massive pink bill, and typical bill-pouch under lower mandible.

Bill in male in breeding season turns reddish-pink. Legs short and strong. An unmistakable bird. **DISTRIBUTION** Found right around mainland and Tas., except central-western deserts in SA, NT and WA. **HABITS AND HABITAT** Very adaptable bird, found on fresh, brackish and saline water from inland to the sea. Single successful chick in nests is usually the oldest and undertakes infanticide, killing subsequent chicks. Birds are capable of moving inland in large numbers in response to rain, with large breeding events occurring on Lake Eyre (SA) and its surrounds once or twice each decade.

Black-necked Stork

■ *Ephippiorhynchus asiaticus*
1.1–1.4m; WS 1.9–2.2m

DESCRIPTION Easily recognized, tall, black-and-white stork with glossy black head and neck (which can appear greenish in certain light), large, pointed black bill, and very long, pinkish-red legs. Male has black eyes, female yellow eyes. DISTRIBUTION Found mainly across northern Australia, from Pilbara region of WA around central NSW coast. Can occur a long way inland at times, although these records tend to be more of vagrant than of regular movements. HABITS AND HABITAT Black-necked Storks, also known as Jabiru, are stunning large birds inhabiting a variety of wetlands, from freshwater lagoons and dams, to coastal mudflats and mangroves. Usually seen singly or in pairs, they forage in their favoured habitat by walking and sweeping the bill from side to side, probing or stabbing prey like fish and crustaceans.

Australasian Bittern

■ *Botaurus poiciloptilus* 66–76cm

DESCRIPTION Large, very stocky-looking bittern, with mottled light brown, dark brown and buff plumage. Throat slightly paler, and best seen when bird is alerted to an observer's presence, when it freezes with its head pointed to the sky. DISTRIBUTION Found in south-east mainland and Tas., and also in WA. Seasonal movements appear to exist for birds in Riverina district of NSW. HABITS AND HABITAT An extremely cryptic species that is really hard to see. Stalks prey in favoured habitat of reed-bed wetlands, snatching at frogs and fish from elevated reeds and other substrates. Also inhabits other wetlands and rice fields. Best way to find birds is to listen for their booming calls in breeding season.

Black Bittern

▪ *Dupetor flavicollis* 54–66cm

DESCRIPTION Dark grey-brown bittern. Both sexes have heavy black-and-white streaking down throat bordered by bold yellow lines down each side. Female smaller than male, with slightly duller plumage.
DISTRIBUTION Found on coasts around northern and eastern Australia, from Pilbara (WA) to far east Gippsland (Vic.), where sightings appear to be getting more reliable. Presumed extinct in south-west WA.
HABITS AND HABITAT Tends to occur in association with tidal rivers and estuaries, where it is found in sheltered trees like melaleucas and paperbarks, but can be a long way from the sea, too. Almost always found singly or in pairs, skulking away if disturbed, or freezing with head held up to the sky.

White-necked Heron

▪ *Ardea pacifica* 76–106cm

DESCRIPTION Breeding birds have all-white head and neck, which terminates abruptly at breast and back where plumage is slaty black-grey. Maroon plumes also appear on back during breeding season. In non-breeding season foreneck spotted black, and maroon plumes are moulted out. **DISTRIBUTION** Widely distributed across mainland Australia and Tas., although largely absent from Great Victoria Desert and its surrounds. **HABITS AND HABITAT** Found singly, in pairs and in loose companies of hundreds of birds (particularly in Top End). Stalks prey in shallow freshwater habitats. Favoured areas include ephemeral wetlands, irrigated pastures and rice crops, farm dams and roadside puddles. In flight white spots at leading edge of wing are reminiscent of lights on wings of a landing plane.

Great Egret ■ *Ardea alba* 83–103cm

DESCRIPTION Large, long-necked, all-white egret. Long, mostly yellow bill, which can be black at tip. Yellow facial skin and gape extend back behind eye, contrasting with facial skin of **Intermediate Egret** *A. intermedia*, which stops at eye. Skin around face flushes greenish in breeding condition, and nuptial plumes develop on back. Legs blackish. **DISTRIBUTION** Widespread across mainland and Tas., although largely absent from central-western desert region. **HABITS AND HABITAT** Utilizes wide array of aquatic habitats, from freshwater to estuarine and saline ones. Most often found singly or in small groups. Forages for its favoured prey of fish, frogs and aquatic insects by stalking slowly before plunging beak into water to snatch at items. Often seen roosting with other egret species.

Great-billed Heron
■ *Ardea sumatrana* 1–1.1m

DESCRIPTION Tall, heavily built, predominantly dark grey-brown heron. Dark bill and legs, and grey-brown upperparts and buff-brown underparts. In breeding condition develops silvery-black plumes on lower throat, crown and back. **DISTRIBUTION** Restricted to coastal fringe of northern and north-eastern Australia, from near Derby (WA) around Top End to south coast of Qld. **HABITS AND HABITAT** Shy and secretive species, often very hard to find. Almost always found singly. Forages on muddy shores, mudflats and shallow water of estuaries and terrestrial wetlands, searching for fish and crustaceans. When not foraging often found roosting on branches of mangroves and similar coastal vegetation.

White-faced Heron
■ *Egretta novaehollandiae* 66–68cm

DESCRIPTION Smallish heron that is nearly uniformly grey all over. Face and under lower mandible white, beak blackish and long yellow legs. When breeding, plumes on back and breast become more prominent. **DISTRIBUTION** Widespread across mainland and Tas.; absent from central south-west desert regions. **HABITS AND HABITAT** The heron most likely to be seen in Australia. Wades in shallow water stalking favoured prey of fish, crustaceans and invertebrates. Uses a variety of habitats, including freshwater wetlands, flooded grassland and farm dams, as well as saline areas like estuarine mudflats, saltmarsh, beaches and lagoons. Also comfortable in urban areas and often seen foraging in roadside puddles and lakes within housing estates.

Straw-necked Ibis ■ *Threskiornis spinicollis* 60–70cm

DESCRIPTION Unmistakable ibis. Back, wings and band of neck/breast iridescent greenish-black, with white underbody, tail and collar around top of neck. Typical bare-skinned head and long, downcurved bill, and diagnostic plume of dirty yellow, 'straw-like' feathers on foreneck. **DISTRIBUTION** Distributed across northern and eastern mainland, and south-west WA, and non-breeding visitor to eastern Tas. Main centre of occurrence is broad Murray-Darling river basin of eastern Australia. **HABITS AND HABITAT** Often forms large flocks, and can be seen flying in classic 'V' formations across the sky or wheeling high into the sky. Grassland and irrigated pastures are the preferred habitat, where it can descend in large numbers to feed on soil invertebrates. Also uses freshwater wetlands, sports ovals and sewage ponds.

Yellow-billed Spoonbill ▪ *Platalea flavipes* 76–92cm

DESCRIPTION Unmistakable at all ages. Wholly white with long, pale yellow, 'spoon-like' bill and yellow legs. In breeding condition develops patch of yellowish plumes on breast and several long black plumes in wings. **DISTRIBUTION** Occurs widely across northern and eastern mainland, and south-west WA, and more casually in eastern Tas. Breeding usually concentrated in lower Murray-Darling river basin of eastern Australia. **HABITS AND HABITAT** Foraging behaviour almost as distinctive as appearance, with birds wading through shallow waters swishing their bills from side to side in search of prey like fish and aquatic invertebrates. Usually seen in small numbers, and often in association with **Royal Spoonbill** *P. regia*. Favours small freshwater wetlands, farm dams, and modified environments like irrigation channels and rice fields.

Black-shouldered Kite ▪ *Elanus axillaris* 35cm; WS 80–100cm

DESCRIPTION Smallish, falcon-like kite, with white head, underparts and tail. Back and most of wings light grey, with large patch of black feathers on shoulder. Yellow legs and red eye. Small amount of black feathering around eye. Underwing white with black primary feathers, contrasting with related **Letter-winged Kite** *E. scriptus*, which has black 'M' underwing pattern. **DISTRIBUTION** Widespread across Australia, although considered vagrant to Tas. Favours coastal regions, but also irruptive in response to rodent plagues. **HABITS AND HABITAT** One of only a handful of raptors in Australia able to 'hover', and the most proficient at it. Faces into the wind and hold wings outstretched, or flaps them to hold position while it searches for signs of mice and rats. Generally favours grassy habitats, either natural or pastures, but also utilizes heathland and saltbush.

Square-tailed Kite
■ *Lophoictinia isura* 50–55cm; WS 1.30–1.45m

DESCRIPTION Reddish-brown kite with dark upperwings and back, and grey area of shoulder on folded wing. Face and head white with faint black streaking. Underwing pattern shows broad black-and-white barring on well-spread primary feathers, with white patch at base of primaries contrasting with rufous of wing near body. Well-fanned square tail in flight. **DISTRIBUTION** Range covers northern and eastern Australian mainland, and south-west WA. An uncommon bird, migratory at southern end of its range, where present only in spring–summer. **HABITS AND HABITAT** Almost a specialist bird predator of songbirds, floating through or over foliage in search of unsuspecting birds or nests to rob. Also takes insects, small mammals and lizards. Found in a variety of habitats, including temperate and subtropical woodland, heathland, rainforests and well-timbered watercourses.

Black-breasted Buzzard ■ *Hamirostra melanosternon* 50–60cm; WS 1.45–1.55m

DESCRIPTION Large raptor, with dark upperparts broken up by shaggy, rufous-brown nape and upper neck, and dark brown margins on some wing-coverts. Breast is black as

its name suggests. In flight shows large, broad dark wings with solid white band at base of well-spread black primaries. **DISTRIBUTION** Core range includes eastern inland, northern and western Australia. Can disperse to more southern latitudes during summer months. **HABITS AND HABITAT** Renowned for tool-using ability – able to break into eggs of ground-nesting birds by 'throwing' stone from beak until egg is broken. Mainly forages on rabbits, lizards, small–medium songbirds, nestlings and carrion. Generally prefers timbered watercourses and associated floodplains, and surrounding savannah grassland.

White-bellied Sea-Eagle ▪ *Haliaeetus leucogaster* 75–85cm; WS 1.8–2.2m

DESCRIPTION Very large, grey-and-white eagle. Head, neck, underbody and tail white except for dark base; back and upperwing dark grey. Soars on long, upswept wings that from underneath are white at front and grey-black at rear. Juveniles and immatures have varying amounts of brown plumage, decreasing with age. **DISTRIBUTION** Found around whole coast and islands of mainland and Tas., and also well inland on suitable waterbodies of northern, eastern and south-eastern Australia. **HABITS AND HABITAT** Favours marine habitats around coast, and drawn into inland areas by large rivers or permanent lakes. Range of each resident pair is very large, and territorial disputes involve spectacular wheeling to the ground of birds locked at the talons. Often seen snatching fish from the water's surface, and also opportunistically takes birds, mammals, reptiles and carrion.

Whistling Kite
▪ *Haliastur sphenurus* 50–60cm; WS 1.2–1.4m

DESCRIPTION Slightly scruffy-looking, pale brown kite. Plumage rufous-brown over head, neck and body, interspersed with pale streaking. Wings darker brown with feathers fringed buff-brown. In flight soars with level but arched wings, and has very long, rounded tail. Underwing dark with brown 'M' pattern; well-spread primaries. **DISTRIBUTION** Widespread across mainland, although not often seen in driest inland deserts. Partly migratory, and only a casual visitor to Tas. **HABITS AND HABITAT** Found in wooded habitats, open areas such as grassland, and marine and terrestrial wetlands. Call a common sound along inland rivers and wetlands, with whistle carrying over a great distance. Typically feeds on carrion in non-breeding season, which renders it susceptible to vehicle strikes, but when breeding favours live food like rabbits and other mammals, birds and lizards.

Brown Goshawk

■ *Accipiter fasciatus* 40–55cm; WS 75–95cm

DESCRIPTION Beautifully patterned plumage, with adults having reddish-rufous chest and belly barred finely with white, and reddish nape-band. Head and wings grey-brown. Eyes and legs yellow, and 'eyebrow' heavy, giving impression that bird is frowning. Tail long and round, contrasting with similar but smaller **Collared Sparrowhawk** A. *cirrocephalus*, which has shorter, square-cut tail. Female larger than male. **DISTRIBUTION** Distributed across whole mainland and Tas., although forests of south-east mainland and Tas. are the stronghold. Several subspecies. **HABITS AND HABITAT** Serious songbird predator, cruising through woodland and forest habitats in search of unsuspecting prey. Ambushes foraging birds within foliage, and also capable of chasing birds in adjacent open areas. Also takes nestlings, small mammals, reptiles and invertebrates.

Spotted Harrier ■ *Circus assimilis* 50–62cm; WS 1.2–1.5m

DESCRIPTION Large, slim-bodied raptor, blue-grey above and rich rufous underneath. Underbody and wings spotted white, and black-and-white barring on underside of black-

tipped primaries. Typical harriers' owl-like face, and long yellow legs. **DISTRIBUTION** Found across entire mainland throughout the year, although tends to visit southern Australia more in summer and autumn, and goes as far as northern Australia in the dry season. Vagrant to Tas. **HABITS AND HABITAT** Generally bird of arid and semi-arid regions. Searches by quartering (long, low glides) for favoured rodent prey over rangeland, farm paddocks and native grassland. Also seen in mallee, spinifex and saltbush areas. Can be irruptive in response to rodent plagues, but also feeds on ground birds like quail, songlarks and pipits, and some reptiles.

Wedge-tailed Eagle
■ *Aquila audax* 85–105cm; WS 1.8–2.3m

DESCRIPTION The largest of Australia's raptors. Large, light brown to dark brown bird, with plumage darkening with age. Well-feathered legs. Impressive sight in flight, soaring on thermals with upheld wings, and large diagnostic wedge- or diamond-shaped tail fanned out. **DISTRIBUTION** Found right across mainland and Tas., where local subspecies is endangered. **HABITS AND HABITAT** Impressive bird with enormous home range in breeding season of hundreds of hectares. Birds pair for life, and raise a clutch every 2–3 years. Young stay with parents for up to two years. Nests reused and improved each year, and can become massive. Utilizes all manner of habitats, except most densely populated cities, and favours mammals such as rabbits, and carrion, as prey.

Nankeen Kestrel ■ *Falco cenchroides* 30–35cm; WS 60–80cm

DESCRIPTION Smallest falcon in Australia, with distinctive plumage. Back and wing-coverts chestnut flecked with black; underside whitish. Black 'teardrop' mark on face. Male has grey head and tail, while female has chestnut head and chestnut tail with black subterminal band. **DISTRIBUTION** Widespread across mainland and Tas., although uncommon in latter. Seem to show seasonal migratory movements. **HABITS AND HABITAT** Very skilled at hovering, and can appear motionless when facing into the wind as it hovers to search for small rodents. Uses habitats from open farmland and rangeland to woodland, and has also adapted to urban areas, often nesting in large buildings. Also consumes large insects, at times in high country converging to gorge on large moths as they emerge at dusk.

Brown Falcon ■ *Falco berigora* 40–50cm; WS 90–120cm

DESCRIPTION Medium-sized falcon, with plumage ranging from almost wholly dark sooty-brown to birds with reddish-brown backs and pale undersides. Variable colour thought to

be different morphs, but now also thought to represent differences due to age and sex. Double 'teardrop' mark on cheek and several other plumage characteristics are constant in the species. **DISTRIBUTION** Found right across Australian mainland, Tas. and offshore islands. **HABITS AND HABITAT** Very opportunistic hunter, taking prey as diverse as rabbits, birds as big as galahs, snakes and insects. Despite being quite 'broad winged' for a falcon, deceptively agile. Found across both open and wooded habitats, but generally avoids thick forest. Usually solitary or in pairs. Quite vocal and often gives raucous cackling calls when approaching a mate.

Black Falcon ■ *Falco subniger* 45–55cm; WS 95–115cm

DESCRIPTION The biggest of Australia's falcons – large, slim dark bird. Plumage almost uniformly sooty-dark brown to black, with very faint pale patch in cheek. Bill dark blue-

grey, and bare parts of face have blue tinge. Old birds develop whitish, black-streaked throat. In flight they show long, pointed wings and long, rounded tail. **DISTRIBUTION** Distributed around inland parts of eastern and northern mainland, dispersing to coastal regions of south in summer–early autumn, and northern coast during dry season. **HABITS AND HABITAT** Skilled aerial predator of small birds, and often perches around waterholes to prey on birds such as pigeons, finches and parrots as they come to drink. Tends to favour open country of rangeland and grassland, but also uses timbered rivers and creeks, farmland and wetlands.

Brolga ■ *Grus rubicunda* 77–134cm

DESCRIPTION Elegant, tall, light grey crane with red-banded head. Under chin there is a distinctive blackish dewlap, which is bigger in male, which also tends to be 20–30cm taller than female. Can be confused with **Sarus Crane** G. *antigone*, but this bird lacks dewlap of Brolga, and its red head plumage continues down neck. **DISTRIBUTION** Found across northern and eastern Australia, although in east tends to be concentrated in southern NSW and western Vic. Occurs in arid areas of south-west Qld and north-east SA when Channel Country floods. **HABITS AND HABITAT** Famed for mating ritual, where males dance and trumpet, wings outstretched and head raised, in front of a female which trumpets in response. Found in pairs to large post-breeding 'flocks'. Uses wide variety of habitats, including shallow freshwater wetlands, irrigated pastures, crops and ploughed paddocks, stubble and grassland.

Buff-banded Rail ■ *Gallirallus philippensis* 30–33cm

DESCRIPTION Beautiful tall rail with reddish bill, white eyebrows, and chestnut cheeks and nape. Underparts barred black and white apart from distinctive buff line across breast; back dark brown with faint white spots. **DISTRIBUTION** Found around broad coastal fringe of mainland, apart from central northern and southern coasts. Occurs further inland, particularly in south-east Australia, when conditions are favourable. Vagrant to Tas. **HABITS AND HABITAT** Found in vegetation around aquatic habitats like freshwater wetlands, estuaries and beaches. Generally wary but in some areas has adapted to human activity, and can become quite bold and confiding. Usually most active at dawn and dusk, foraging around muddy margins or in open water close to vegetation.

Baillon's Crake ■ *Porzana pusilla* 15–18cm

DESCRIPTION Smallest Australian rail, with pale blue-grey face and underbody that becomes barred on belly and continues onto cocked undertail. Upperparts pale brown with broad black streaking. Red eyes, and dark green bill and legs. **DISTRIBUTION** Occurs from

coastal WA and northern Australia, where it is uncommon, into eastern Australia, including Channel Country and its stronghold of Murray-Darling Basin. Also north and east coast of Tas., although uncommon there. **HABITS AND HABITAT** Tends to favour well-vegetated freshwater shallow wetlands, either permanent or ephemeral, but can also tolerate brackish and saline environments. Found singly or in pairs. Secretive and nervous, often dashing for cover without apparent reason. Foraging usually done on floating vegetation, and aquatic insects make up bulk of diet.

Australian Spotted Crake ■ *Porzana fluminea* 19–23cm

DESCRIPTION Face, neck and breast dark blue-grey, and bold black barring on white belly and flanks. Underside of cocked tail mostly white. Upperparts dull brown with fine white spotting and streaking. Eye red, and beak yellow with broad reddish cere. Legs olive-

green. **DISTRIBUTION** Uncommon around coastal WA and northern Australia, and most commonly found in south-eastern Australia, including in its stronghold of Murray-Darling Basin. Also occurs on north-east coast of Tas. **HABITS AND HABITAT** Usually occurs in singles or pairs in vegetated margins of permanent or ephemeral wetlands. Favours freshwater environments, where it prefers to wade in shallow water or forage in muddy margins, but also able to utilize brackish and saline habitat. Generally the boldest of the crakes found in Australia, and often forages confidingly out in the open when being observed.

Black-tailed Native-hen ■ *Tribonyx ventralis* 30–38cm

DESCRIPTION Stocky-looking bird with dull olive-brown upperparts, and blue-grey breast and underside. Broad white striations on sides of breast, yellow eye and yellow-green bill with red base to lower mandible. Legs bright orange and dark tail held erect.
DISTRIBUTION Mostly found in large drainage basins of south-east and central-east of mainland, as well as along WA coast from south-west to around Pilbara.
HABITS AND HABITAT Colloquially known by many birders as the 'turbo chook', it can move extremely quickly when disturbed. Generally found in small parties, and may move in response to substantial rain or flood events, congregating in large numbers. Favours permanent and ephemeral freshwater wetlands, especially those containing lignum, but can also be found in open woodland as well as crops and gardens.

Dusky Moorhen ■ *Gallinula tenebrosa* 35–40cm

DESCRIPTION Medium-sized waterhen. Dull brown on back and wings, and purplish-black on head, neck and underbody. Bright red bill tipped with yellow, with red extending into red bill-shield on face. White undertail-coverts. **DISTRIBUTION** Found mostly along eastern coast and inland to central NSW, Qld and northern Vic., as well as in northern Tas. and islands of Bass Strait, south-east SA and coastal south-west WA. **HABITS AND**

HABITAT Found most often in loose flocks and congregations in association with **Eurasian Coot** *Fulica atra* and **Purple Swamphen** *Porphyrio porphyrio*. Fleet-footed bird capable of running quickly on land, and also very adept at swimming. Favours permanent and ephemeral freshwater wetlands, although at times tolerates and uses brackish to saline areas, with vegetable matter, seeds, molluscs and invertebrates among the favoured foods.

Australian Bustard

■ *Ardeotis australis* 80–120cm; male larger than female

DESCRIPTION Unmistakable large, solid bird of inland plains. Long, creamy-white neck, throat, breast and underside; back chestnut with fine barring. Leading edge of folded wing has black feathers with white margins. Both sexes have black crown, and male has black band across breast. **DISTRIBUTION** Mainly found across inland and northern Australia, and uncommon and more irregular visitor to southern and south-eastern mainland regions. Absent from Tas. **HABITS AND HABITAT** Found singly, in pairs and in loose flocks. Wanders with bill usually held aloft through favoured habitat like grassland, shrubland and very open inland woodland in search of prey. During breeding season male develops large, loose vocal sac that can touch the ground, which it bellows from to attract females.

Bush Stone-curlew ■ *Burhinus grallarius* 54–59cm

DESCRIPTION Large, tall, ground-dwelling bird. Generally dark grey streaked black above, and white streaked black underneath. Wings have large white patches, particularly

visible in flight. Head rounded, large yellow eye, and white forehead and eyebrow. A rufous morph has rufous plumage in throat and wings. **DISTRIBUTION** Spread around most of mainland except western NSW, SA and arid areas of WA. Common in parts of northern Australia, but distribution is fragmented and contracting in south-east. Absent from Tas. **HABITS AND HABITAT** Birds of southern Australia severely impacted by predation from introduced Red Fox, with local extinctions in many districts. Perhaps as a result, southern Australian birds are very cryptic and wary, often lying motionless on the ground; birds in northern Australia are more approachable. Presence often given away by calls at dusk, which are very eerie and far carrying and often undertaken by multiple birds. Usually nocturnal.

Beach Stone-curlew ▪ *Esacus giganteus* 54–56cm

DESCRIPTION Slightly comical-looking bird, due to extremely large black beak and yellow eye set under white eyebrow in dark face. Dark grey-brown above and pale below, with dark line in folded wing. In flight shows large white panels near end of wing. **DISTRIBUTION** Found around coast and associated coastal islands of 'northern' Australia, from southern Pilbara coast of WA, around to Nambucca Heads, NSW. Infrequent vagrant to Vic., as far as Otway Coast. Absent from SA and Tas. **HABITS AND HABITAT** Usually found alone or in pairs. Nocturnal species, but can be active at dusk and dawn. Prefers isolated and uninhabited beaches, particularly those with areas of mudflat, mangrove or tidal pan that are exposed for foraging at low tide.

Australian Pied Oystercatcher ▪ *Haematopus longirostris* 42–50cm

DESCRIPTION Handsome pied bird. Black upperbody, head and foreneck, and clean white underside. Bright orange-red bill, orange eye and eye-ring, and pink legs. Easily distinguished from **Sooty Oystercatcher** *H. fuliginosus*, whose plumage is wholly black. **DISTRIBUTION** Found in suitable habitat around coast of mainland and Tas. Also some offshore islands. **HABITS AND HABITAT** Found singly and in small flocks. Forages on intertidal mudflats, sandy beaches and sandbanks, looking for molluscs, invertebrates, crabs and occasionally small fish. Nest a simple sand scrape usually in dune or similar area above high-tide mark, and often concealed among flotsam or vegetation. Like other beach-nesting species, subject to high levels of egg predation by ravens and introduced Red Fox.

Pied Stilt ■ *Himantopus leucocephalus* 33–37cm

DESCRIPTION Tall, elegant wading bird. White head, foreneck, rear collar and underside, and black wings and nape. Bill long, narrow and black. Long pink legs. Juveniles similar to

adults, but have smutty dark head that lacks black nape. **DISTRIBUTION** Widespread across all of mainland and Tas. in suitable habitats, apart from driest deserts of eastern WA and western SA. **HABITS AND HABITAT** Very vocal species, and its dog-like *yap, yap* calls are one of the most common sounds within its favoured shallow freshwater wetlands. Types of wetland utilized include swamps, sewage-treatment ponds, saltworks, billabongs and lakes, particularly those with short emergent vegetation. Often found in association with Red-necked Avocets (see below).

Red-necked Avocet ■ *Recurvirostra novaehollandiae* 40–48cm

DESCRIPTION Unmistakable species with rusty-red head, diagnostic long, narrow upturned black bill, and white body with black-and-white wings. Legs slatey blue-grey.

DISTRIBUTION Found predominantly across southern Australia, although generally absent from Tas. Favoured breeding areas are lower Murray-Darling Basin, ephemeral wetlands of Lake Eyre region (SA), and inland south-west WA. Rarely in Tas. **HABITS AND HABITAT** One of only a handful of species in Australia that forages by sweeping slightly open bill from side to side through the water or mud as it walks, picking out prey items in the process. Prey includes insects and crustaceans. Able to use a variety of freshwater, brackish and saline habitats, although saline areas are generally favoured. These particularly include evaporating inland wetlands that increase in salinity as they dry out.

Banded Stilt ■ *Cladorhynchus leucocephalus* 35–43cm

DESCRIPTION Graceful but solid bird with all-white head and body, predominantly black wings and long, narrow black bill. Adults in breeding condition develop bold rusty-coloured chest-patch that narrows to stripe on belly, which fades or disappears in non-breeding birds and is absent in juveniles. **DISTRIBUTION** Predominantly found in favoured habitat in western Vic., coastal and Lake Eyre basin of SA, and inland south-west WA (particularly on salt lakes). Vagrant to Tas. **HABITS AND HABITAT** The quintessential bird of saline wetlands, with large concentrations gathering in areas such as saltworks, shallow saltwater lakes and tidal mudflats. Also occasionally utilzes freshwater wetlands. Feeds by probing the substrate and pecking on the surface for prey such as crustaceans, molluscs and insects, although some vegetation is also taken.

Red-capped Plover ■ *Charadrius ruficapillus* 14–16cm

DESCRIPTION Small plover, white underneath with grey-brown back and wings. Male has distinctive reddish crown and nape that is bordered by very fine black line, and black line through eye. Female's head duller with fainter black markings. **DISTRIBUTION** Widespread across mainland and Tas., and coastal islands. **HABITS AND HABITAT** Shorebird that favours brackish and saline wetlands. Common in suitable habitats such as saltmarsh, commercial saltworks, broad sandy beaches, tidal mudflats and drying inland wetlands. Nest a shallow scrape in the sand, or occasionally on shell/shingle or bare ground; if nest is approached too closely birds perform 'broken-wing' distraction display to lead away the observer. Although common, it does suffer from disturbance from human activities, particularly when nesting.

Inland Dotterel
■ *Peltohyas australis* 19–23cm

DESCRIPTION Small, pretty bird with sandy-brown underside, mottled darker brown upperparts, and distinctive black breast-band and belly-stripe. White face and foreneck accentuate dark teardrop mark under each eye. **DISTRIBUTION** Found across inland regions of all mainland states, although only just into Vic. Stronghold is broad area of western NSW, south-west Qld and inland SA. **HABITS AND HABITAT** In many ways an anomaly, this 'wader' has evolved to exist on arid inland plains. It moves nomadically, in response to rainfall and/or changes in vegetation structure. Feeds mainly on seeds, leaves and invertebrates like insects and spiders. Usually found while foraging at night, it is quite gregarious and usually seen in small flocks. Habitats used include gibber plains, claypans and bluebush/saltbush areas.

Black-fronted Dotterel ■ *Elseyornis melanops* 16–18cm

DESCRIPTION Small shorebird with grey-brown upperparts and white underparts broken by broad black breast-band. Broad black line of feathers through face, bordered

above by thin white line. Red eye-ring, black-tipped red beak and pink legs. **DISTRIBUTION** One of Australia's most widespread waders, occurring around mainland and Tas., except driest inland western deserts. Stronghold is south-east and south-west mainland. **HABITS AND HABITAT** Favours shallow freshwater wetlands, particularly those with muddy margins or bottoms. Walks around probing or pecking the mud for prey such as crustaceans, molluscs and insects. Also tolerates brackish environments, and often found in urban wetlands close to human disturbance. Pairs with eggs or chicks feign injury if offspring are under threat, in order to lead away disturbance.

Hooded Plover ■ *Thinornis cucullatus* 19–23cm

DESCRIPTION Unmistakable, medium-sized shorebird with somewhat stocky appearance, and all-black head with red eye-ring and red bill that is tipped black. Black collar marking around rear of neck, grey wings and back, and white underbelly. Dull orange-pink legs.

DISTRIBUTION Found around coasts of eastern SA, Vic., southern NSW and Tas., and on inland salt lakes as well as coast of south-west WA. **HABITS AND HABITAT** In eastern states favours high-energy ocean beaches, where it forages on tide line and associated bays and inlets. In WA also utilizes inland salt lakes as habitat. On ocean beaches tends to nest around tide line and often in higher dunes. Nest a simple scrape on the ground.

Banded Lapwing
■ *Vanellus tricolor* 25–29cm

DESCRIPTION Distinctive endemic lapwing, not easily confused. Back and wings grey-brown, and underbody clean white. Head black apart from broad white eye-stripe, and black continues in 'U' shape down around front of lower breast. Throat all white, bill and eye yellow; small round red wattles above bill. **DISTRIBUTION** Spread across southern half of mainland and Tas., with distribution broadly petering out in line from central Qld coast to Exmouth Gulf, WA. Most common in Riverina (NSW), south-east SA and wheat belt of WA. **HABITS AND HABITAT** When nesting usually occurs only in pairs, but in non-breeding season can be quite gregarious. Nomadic or dispersive. Favours open habitat containing short grass or no grass, particularly recently ploughed paddocks, new crops and/ or stubble. Quite vocal and calls often when danger is present.

Masked Lapwing ▪ *Vanellus miles* 30–37cm

DESCRIPTION Australia's largest lapwing, with diagnostic yellow wattles above and below eye. Cap black in all subspecies, extending down sides of neck on southern-eastern

Australian birds. Grey-brown back, clean white underside and pink legs. Spurs on bend of wing used in threat and to attack approaching danger. **DISTRIBUTION** Found around northern and eastern sections of mainland, and Tas. Most common in south-east Australia. **HABITS AND HABITAT** Highly territorial in breeding season; will take to the wing and aggressively defend nest by regularly flying at intruders with wing spurs revealed. Tolerant of humans and often found nesting on sporting ovals and open spaces; also utilizes freshwater wetlands and saltmarsh. Large flocks form in non-breeding season. Formerly known as Spur-winged Plover.

Plains-wanderer ▪ *Pedionomus torquatus* 15–19cm; female larger than male

DESCRIPTION Sexually dimorphic, with female the more attractive bird. Back and underside brown with dark blackish mottling through plumage; paler on belly. Upper

throat of female black with ornate white spots, bordered below by rich chestnut-coloured breast. Male lacks spotted throat and chestnut breast, and paler over body than female. **DISTRIBUTION** Extremely restricted range, with core area being Riverine Plains of NSW and north-central plains of Vic. Also found in pastoral areas of south-west Qld and north-east SA, although current status there is unclear. **HABITS AND HABITAT** True bird of native grassland; taxonomically unique 'shorebird' that has evolved for life on sparse, grassy plains of inland south-east Australia. Male incubates eggs and broods young. Species best found by spotlight in suitable habitat, but recent declines make it extremely difficult to find.

LEFT Male; RIGHT Female

Comb-crested Jacana

■ *Irediparra gallinacea* 20–27cm;
female larger than male

DESCRIPTION Unmistakable. Glossy deep
pink wattle on top of head, with creamy face
and throat that becomes golden at lower end.
Dark brown back and wings, broad black
breast-band and clean white belly. Long dark
legs, and enormous elongated toes and claws.
DISTRIBUTION Found around coastal eastern
and northern Australia, from Hunter Estuary,
NSW, to Kimberly coast in WA. **HABITS AND
HABITAT** Also known as Lily Trotter and Jesus
Bird. Elongated feet allow birds to walk on tops
of water-lilies and other vegetation floating on
water. In flight long legs and toes trail far behind
birds. Favours freshwater wetlands and lagoons,
where it forages for prey such as aquatic insects
and seeds. Lays 3–4 eggs in nest of floating
vegetation.

Australian Painted Snipe ■ *Rostratula australis* 24–30cm

DESCRIPTION Stunning bird, dark above and white below with white 'saddle' marking
over shoulder. White comma-shaped patch of feathers around each eye, golden-yellow
line on crown and long, slightly downcurved
bill. Female has glossy rufous head and neck,
and finely barred wings. Male has browner
neck and wings with yellow-and-buff spots.
DISTRIBUTION Found anywhere around
mainland except driest western and southern
deserts, but Murray-Darling basin of south-east
Australia is the stronghold. **HABITS AND
HABITAT** Another Australian endemic shorebird
exhibiting polyandrous mating system, where
female breeds with multiple males in a season,
which incubate and brood the young alone.
Found almost exclusively in shallow freshwater
wetlands with vegetation such as lignum and
other emergent plants that allow concealment.
Also responds to flooding and filling of ephemeral
wetlands by rainfall, travelling long distances to
such suitable habitats.

Latham's Snipe ■ *Gallinago hardwickii* 29–33cm

DESCRIPTION Black cap on top of head. Face has several dark stripes giving appearance of long, whitish supercilium and cheek. Back and wings mottled and streaked in browns, black

and white. Pale underbelly and long, straight bill. **DISTRIBUTION** Summer migrant from Japan; present around east coast of mainland and into Tas.; west as far as Eyre Peninsula in SA. **HABITS AND HABITAT** Found in a range of freshwater wetlands, especially those with low cover. Occurs singly or in small groups, although can be found in large flocks on suitable wetlands. Very shy and secretive. Explodes from cover if disturbed and flies away rapidly, usually giving rasping *kzeck* call. Mainly crepuscular.

Bar-tailed Godwit ■ *Limosa lapponica* 37–39cm

DESCRIPTION Medium-sized shorebird with long, slightly upturned pink bill with dark tip. Usually present in non-breeding plumage; upperparts grey-brown with some fine

streaking, neck lighter and streaked, pale unstreaked belly, and tail diagnostically barred blackish and white. Can be confused with **Black-tailed Godwit** *L. limosa*. **DISTRIBUTION** Summer migrant from northern hemisphere; can be found anywhere around coast of mainland and Tas., except section of SA–WA coastline. Highest numbers in northern Australia. **HABITS AND HABITAT** Gregarious; mostly in small flocks but at times thousands can be present at a site. Favours coastal tidal mudflats and estuaries, but also sandy beaches and sewage farms. Forages by probing in soft mud for invertebrates, which it senses with bill-tip. Tall bird, so able to stand in deeper water than most stints and sandpipers.

Far Eastern Curlew ■ *Numenius madagascariensis* 60–66cm

DESCRIPTION Largest shorebird in Australia. Bulky, with long neck and extremely long, downcurved bill. Plumage overall sandy-brown with heavy dark brown streaking or mottling; pale belly, also streaked brown. **DISTRIBUTION** Summer migrant from northern hemisphere; found anywhere around coast of mainland and Tas., except section of SA–WA coastline. Highest numbers in northwest Australia. **HABITS AND HABITAT** Impressive bird, usually found singly or in small flocks. Considered a coastal species, using beaches, estuaries and saltmarsh to forage for prey like crabs and molluscs. Tends to be very wary and hard to approach, and roosts separately from other shorebirds. Can be found in large flocks in some areas, such as north-west WA.

Grey-tailed Tattler ■ *Tringa brevipes* 25cm

DESCRIPTION Elegant medium-sized shorebird, present in non-breeding plumage in Australia. Grey above, and with pale grey breast fading to pale belly and undertail. Dark cap and eye-stripe around white eyebrow; bill straight and yellowish at base. Easily confused with less common **Wandering Tattler** *T. incana*. **DISTRIBUTION** Summer migrant from Siberia; found anywhere around coast of mainland and Tas. Highest numbers in northern Australia; less common in south. **HABITS AND HABITAT** Usually found in small flocks, but also occurs singly. Utilizes sheltered coastal areas with reefs, rock platforms or intertidal mudflats. Readily mixes with other shorebirds when foraging and roosting. Like other *Tringa* species, walks along foraging with regular pauses, during which it bobs its tail up and down.

Ruddy Turnstone ■ *Arenaria interpres* 22–24cm

DESCRIPTION Unmistakable. Usually found in Australia in non-breeding plumage; short, pointed dark bill, white underside, and white, black and rufous tortoiseshell pattern

on upperparts. Short orange legs. **DISTRIBUTION** Summer migrant from northern hemisphere; can be found anywhere around coast of mainland and Tas., and occasionally inland. Some birds overwinter in Australia. **HABITS AND HABITAT** Well known for distinctive foraging behaviour of turning rocks, shells and seaweed to search for prey. Favours rock platforms and outcrops in association with shallow tidal pools. Usually found in small, loose flocks, and quite tolerant of close approach. Happily intermingles with other shorebirds and terns at roost sites.

Red-necked Stint ■ *Calidris ruficollis* 13–16cm

DESCRIPTION Present in Australia mostly in non-breeding plumage. Small bird with grey back and wings, white underside and small black bill. Sides of neck have grey-brown wash with small dark notches; white

supercilium. **DISTRIBUTION** Summer migrant from northern hemisphere. Found in suitable habitat across all of mainland and Tas., except central Australia. **HABITS AND HABITAT** Remarkable species when return migration to Siberia is considered in light of bird that weighs just 25g. One of the most numerous shorebirds, and often found in the hundreds in mixed flocks with other small and medium-sized shorebirds (such as sandpipers). Prefers coastal mudflats, but also occurs on range of freshwater and brackish wetlands like sewage ponds and saltworks. Walks with hunched shoulders in search of prey.

Sharp-tailed Sandpiper ■ *Calidris acuminata* 17–22cm

DESCRIPTION Medium-sized shorebird with portly shape; usually in non-breeding plumage. Small, flat head with rufous cap and dark eye-stripe giving appearance of white eyebrow. Upperparts, neck and throat grey-brown with dark pointed centres; fades gradually to clear pale belly. Slightly downcurved bill. **DISTRIBUTION** Summer migrant from northern hemisphere; found in suitable habitats across all of mainland and Tas., except central Australia. **HABITS AND HABITAT** One of the most widespread shorebirds in Australia. Typically feeds and roosts with Curlew Sandpipers (see below) and Red-necked Stints (see opposite). Found in range of coastal and inland habitats, fresh, brackish and tidal. Gregarious and often occurs in large, dense flocks of thousands of birds. Similar to rarer **Pectoral Sandpiper** C. *melanotos*.

Curlew Sandpiper ■ *Calidris ferruginea* 18–23cm

DESCRIPTION Medium-sized shorebird seen in Australia in non-breeding plumage; white supercilium on plain dark grey head and upperparts, and grey sides of neck. Faintly streaked, cloudy breast and clean white underparts. Distinctive long, somewhat evenly downcurved bill that is very thick at base. **DISTRIBUTION** Summer migrant from northern hemisphere; found in suitable habitats across all of mainland and Tas., except central-west Australia. **HABITS AND HABITAT** Widespread shorebird in Australia, but has undergone severe declines in previous decade due to changes to flyway stopover points. Associates readily with other small and medium-sized waders. Utilizes variety of coastal and inland habitats, fresh, brackish and tidal. Begins moulting to red-rufous breeding plumage in late summer.

Painted Button-quail ■ *Turnix varius* 17–19cm

DESCRIPTION Generally grey face and underparts with white flecks; upperparts generally rufous-brown with black blotches and some fine white streaking. The only button-quail with

a red eye. Sexes are similar, with male being duller. **DISTRIBUTION** Found coastally and just inland of Great Dividing Range of eastern states; also into south-east SA and south-west WA. **HABITS AND HABITAT** Seen singly, in pairs or in small coveys on the ground, where it forages on items such as insects and seeds. Most active at dusk and dawn, but can be seen during the day. At dusk often heard giving loud *oom* calls, which can carry far. Found in forests and woodland, particularly those with a relatively heavy canopy and understorey, and leaf litter on the ground.

Little Button-quail ■ *Turnix velox* 12–16cm; female larger than male

DESCRIPTION Small button-quail with heavy blue-grey bill and pale eyes. Chestnut-brown upperparts barred and streaked variously black and silvery-white. Pale underside

provides diagnostic white flanks in flight. Female's head and neck plainer pinkish-brown than male's, and lightly scalloped. Male is darker, and heavily patterned on head and rear neck. **DISTRIBUTION** Widely but patchily distributed across mainland, except coastal far northern Australia and sparse south-east of Great Dividing Range; absent from Tas. **HABITS AND HABITAT** Favours grassland and open woodland of temperate and tropical areas, particularly arid and semi-arid habitats. Irruptive and nomadic. Found mostly on inland grassy plains and dunefields, and around escarpments, where it is usually seen singly or in pairs. Runs like a mouse when disturbed; occasionally squats before eventual flight, which takes it low and far from danger.

Australian Pratincole
▪ *Stiltia isabella* 19–24cm

DESCRIPTION Elegant medium-sized, long-legged bird. Upperparts fawn-brown, pale face with dark eye-stripe, and pale underside that develops chestnut-black bar on belly when breeding. Long, pointed dark wings in flight and dark-tipped red bill. **DISTRIBUTION** Found around northern and inland mainland, south to coastal SA. Seasonal migrant; further south during summer only. **HABITS AND HABITAT** Stands tall and upright. Highly gregarious, and mostly found on arid inland on open grassland, lightly wooded plains, gibber plains and associated areas. Often tame and allows close approach. Feeds readily on the ground, but also adept at hawking insects in flight. Mainly crepuscular, loafing during the day.

Brown Skua ▪ *Stercorarius antarcticus* 61–64cm

DESCRIPTION Large, heavy-bodied brown bird similar in appearance to gulls, with large, hook-tipped black beak. In flight, pale base to flight feathers is conspicuous.

DISTRIBUTION Breeds in Antarctica and sub-Antarctic islands over summer, and migrates regularly to Australian waters during winter. Seen in southern Australia from WA to southern Qld. **HABITS AND HABITAT** Pelagic species that often ventures to inshore breaks and sheltered harbours in rough weather. Keen predator of nesting seabirds on breeding grounds; otherwise piratical and often harasses birds bigger than itself. Also known to follow fishing boats in search of discarded offal, and to chase other seabirds until they disgorge food.

Arctic Jaeger ■ *Stercorarius parasiticus* 46–67cm, including tail streamers

DESCRIPTION Two colour morphs that are variable depending on season. Light morph generally pale grey-brown above and buff below, with darker collar and black cap. Scarcer

dark morph more uniformly dark brown, with yellowish collar when breeding. Both morphs have long, pointed tail streamers, absent or reduced when not breeding, and pale bases to outer flight feathers. **DISTRIBUTION** Breeds in Arctic, and oversummers in Australian waters in October–April around east, south and west coasts, including Tas. **HABITS AND HABITAT** Seen singly, in pairs and in small flocks. Favours inshore coastal waters and often seen in large bays and harbours. Parasitic species famed for harassment of terns and gulls until they disgorge food, which they catch in flight; also known to harry shorebirds along intertidal mudflats.

Brown Noddy ■ *Anous stolidus* 40–45cm

DESCRIPTION Largest noddy, a large but slender bird with brown body and wings, black primary feathers giving two-toned look to upperwing in flight, and ashy-white

cap on head cut off abruptly at dark line between bill and eye. **DISTRIBUTION** Mainly off Qld coast, but also central and north-west coast of WA. Irregular down NSW coast. **HABITS AND HABITAT** Gregarious bird of tropical and subtropical waters, roosting and loafing in flocks on rock stacks, beaches and cays. Often found in company with other terns and noddies. Usually forages out to sea in mixed flocks; occasionally inshore. Does not plunge fully like other terns – uses shallow plunges after hovering just above the water's surface.

Fairy Tern ■ *Sternula nereis* 22–27cm

DESCRIPTION Small, delicate tern, with grey back and wings, and white underside and tail. Black cap and nape extends just forwards of eye, but does not reach bill (unlike in similar **Little Tern** S. *albrifrons*). Yellow beak that becomes dark tipped in non-breeding condition. **DISTRIBUTION** Found around southern and western coasts, from southern NSW to north-west WA; also in Tas. and inshore islands **HABITS AND HABITAT** Gregarious coastal species usually foraging over sheltered inshore waters. Breeds on sandbars, spits and sheltered beaches, where it lays clutch of 1–2 eggs in a shallow scrape on the ground. Readily roosts and breeds among other smaller terns and shorebirds like Red-capped Plovers (see p. 51) and Little Terns.

Whiskered Tern ■ *Chlidonias hybrida* 23–25cm

DESCRIPTION Large marsh tern mostly seen in breeding plumage. Black cap extends to eye and down to nape, grey neck, wings and back, and dark sooty-grey underparts. Dark crimson red bill and red legs. Dark bill and legs in non-breeding condition, pale underside and white on crown. **DISTRIBUTION** Summer breeding migrant to Australia, scattered in suitable habitats anywhere except central-western deserts; vagrant to Tas. **HABITS AND HABITAT** Highly gregarious, utilizing a range of freshwater and brackish wetlands, either coastal or inland. Does favour shallow freshwater wetlands with emergent vegetation like grass and reeds; often over wetting and early-growth rice fields. Catches prey by either picking it off surface or plunge diving. Often in company of other terns, especially **White-winged Black Tern** C. *leucopterus*.

Greater Crested Tern ■ *Thalasseus bergii* 40–50cm

DESCRIPTION Very large sea tern with slate-grey back and wings, and white neck and underside. Black cap separated from yellow bill by white stripe; rear of crown in breeding condition shaggy in appearance. Easily confused with **Lesser Crested Tern**

T. bengalensis, which is smaller, paler grey and has orange bill. **DISTRIBUTION** Common around coastal mainland and Tas.; more abundant in southern regions. **HABITS AND HABITAT** Gregarious; usually seen cruising over ocean beaches singly or in small groups, or roosting on jetties, pylons and boats. Associates readily with other terns and gulls. Forages by snatching items from surface on the wing, or plunge diving fully from several metres above the water. Nest a simple scrape in sand or on rock, often on inshore islands in thousands.

Pacific Gull ■ *Larus pacificus* 50–67cm

DESCRIPTION Large gull of coastal southern Australia. Takes up to five years to reach adult plumage of white head, neck and underside, black back, black wings with thin white inner trailing edge, and white tail with black subterminal band. Massive yellow beak tipped

with red, and yellow legs. Immatures off-white, streaked with brown. **DISTRIBUTION** Coastline from Shark Bay in WA to south coast of NSW, Tas. and Bass Strait islands. **HABITS AND HABITAT** Common sight at beaches, bays and harbours of southern Australia. Gregarious, and seen singly, in pairs or in small groups, mostly of immature birds in various plumages. Searches between high-tide line and water's edge in search of prey; known to carry molluscs aloft to drop onto rocks. Often loiters around fishermen for discarded scraps.

Silver Gull ▪ *Chroicocephalus novaehollandiae* 36–44cm

DESCRIPTION Easily identified small gull. White head, neck, underside and tail, and pale grey wings with white-tipped black primary feathers. Bright red bill and legs, and white eyes. Immatures have darker bill and legs, and browner coverts in wing. **DISTRIBUTION** Widespread across mainland and Tas.; more common in southern regions. **HABITS AND HABITAT** Common, tame and well-known gull of coasts and inland areas. Natural habitat a variety of coastal and inland wetlands. Picks prey from surface of water or hawks insects. Adapted and thriving in urban environments, where it scavenges readily around parks, sporting ovals and rubbish dumps. Also a serious nest predator of birds like Banded Stilt (see p. 51), and during significant breeding events.

Palm Cockatoo

▪ *Probosciger aterrimus* 56cm

DESCRIPTION Very large, completely black cockatoo. Enormous black beak, large patch of orange-red bare facial skin, and long, shaggy crest on top of head. Juveniles have yellow-fringed underside feathers and pale cheek-patches. **DISTRIBUTION** Restricted to Cape York Peninsula in Qld, from tip down to region of Edward River and Princess Charlotte Bay. **HABITS AND HABITAT** Usually seen singly, in pairs or in small parties. Noisy and active. Male has a highly specialized courtship display that involves 'drumming', where he bangs a large stick against a hollow tree while calling. Found in tropical woodland and rainforests, especially around ecotones between the two.

Red-tailed Black-Cockatoo

■ *Calyptorhynchus banksii* 55–60cm

DESCRIPTION Five subspecies exist, with the northern one being the largest and the central Australian one the smallest. In all subspecies males are nearly completely glossy black, with a rounded erectile crest and bright red panels in tail. Females have pale yellow spots on upperparts, yellow fringing on underside feathers and orange-yellow barring in tail. **DISTRIBUTION** The five subspecies are found in somewhat discrete areas around Australia, including south-west WA, northern and central Australia, the east coast and the SA–Vic. border. **HABITS AND HABITAT** Usually gregarious and mostly found in large flocks, seen flying noisily across the sky or foraging in tree-tops. Also adept at feeding on the ground, and readily visits waterholes, dams and troughs to drink. With such a wide range, utilizes a variety of eucalypt-dominated forests and woodland, with south-eastern subspecies being dependent on Brown Stringybark for foraging.

Yellow-tailed Black-Cockatoo

■ *Calyptorhynchus funereus* 55–65cm

DESCRIPTION Large, almost wholly black bird, except for yellow cheek-patch and yellow panels on tail. Sexes differ slightly: male has red eye-ring and dark bill, while female has grey eye-ring, light bill and larger yellow check-patches. **DISTRIBUTION** Occurs around south-east Australia from Rockhampton (Qld) to southern Eyre Peninsula and Kangaroo Island (SA), as well as in Tas. **HABITS AND HABITAT** Highly gregarious, and nearly always seen in pairs or small parties. Mainly found in trees, only coming to the ground to drink or feed on fallen pine cones. Utilizes wide variety of habitats, but favours eucalypt rainforest and woodland, banksia woodland and heathland, and pine plantations. Found quite low down at times, extracting larvae from bases of trees.

Carnaby's Black-Cockatoo
■ *Calyptorhynchus latirostris* 54–56cm

DESCRIPTION Large, matte-black cockatoo, with large white cheek-patch and broad white panels in tail. Male has dark bill and red eye-ring. Female has whitish bill and dark eye-ring. Easily confused with **Baudin's Black-Cockatoo** *C. baudinii*, which has a much longer upper mandible among several other subtle differences. **DISTRIBUTION** Restricted to south-west corner of WA, broadly across wheat belt of that district. Frequently seen around Perth. **HABITS AND HABITAT** Also known as Short-billed Black-Cockatoo, the species has declined substantially as a result of historic persecution and continued loss of habitat. Favours eucalypt woodland, particularly that of sandplains around Perth. Also utilizes mallee habitats. Shows regional movements during the year, breeding in drier habitats in spring before moving back to wetter habitats in summer. Highly gregarious.

Gang-gang Cockatoo ■ *Callocephalon fimbriatum* 32–37cm; female larger than male

DESCRIPTION Stunning, unmistakable cockatoo. Male has whitish-edged, slate-grey body feathers and scarlet head and crest, with crest feathers finely filamentous at tips.

Female has dark grey head and crest, and grey underparts with feathers lightly edged with pink and yellow. **DISTRIBUTION** Restricted to coastal and mountain areas of south-eastern mainland, from Hunter Valley (NSW) around to far south-east SA. **HABITS AND HABITAT** Gregarious; usually seen in pairs or small flocks. When foraging and roosting generally quiet and inconspicuous, and usually located by the sounds of falling debris while feeding. In flight makes conspicuous raking calls akin to creaking rusty gate. Found in tall eucalypt forest and woodland, and seasonally can be common in urban parks and gardens.

LEFT Male; RIGHT Female

Major Mitchell's Cockatoo ■ *Lophochroa leadbeateri* 39cm

DESCRIPTION Unmistakable pink-and-white cockatoo. Face, neck and underside bright to pastel pink, and upperparts and wings white. Erectile crest that is generally striped pink

and yellow inside a white border (one subspecies has no yellow in crest). **DISTRIBUTION** Found across semi-arid and arid interior of mainland, from inland Qld, NSW and Vic., through SA and NT, to far coast of WA. **HABITS AND HABITAT** Highly gregarious, usually occurring in pairs or small parties, but at times in larger flocks. Spends most time foraging on the ground or in shrubs and trees, and often found in association with Galahs (see below) and corellas. Habitat used includes timbered watercourses, and surrounding saltbush plains, gibber and grassland. Also found in mallee and native pine woodland.

Galah ■ *Eolophus roseicapillus* 35cm

DESCRIPTION Easily identified bird with rich salmon-pink underparts, and grey wings and back. Also has small cap that is either white or pink. Sexes are alike, although male

has dark brown eyes and female has red eyes. **DISTRIBUTION** One of the most widespread Australian birds, found across entire mainland except waterless deserts and dense forests. Recently self-introduced to Tas. **HABITS AND HABITAT** Highly gregarious, and a quintessential Australian sight as you drive around the countryside. Seen singly, in pairs or in large flocks. Forages mainly on the ground, but also in foliage of trees, and often on roadsides. Noisy and conspicuous, and usually seen in association with Sulphur-crested Cockatoo (see opposite). Uses a variety of timbered habitats, town parks and sporting ovals, generally near water.

Little Corella ▪ *Cacatua sanguinea* 35–40cm

DESCRIPTION Small white cockatoo with shades of yellow in undertail and pink wash to head and throat. Short erectile crest, short whitish bill, blue skin around eye and subtle pink lores. Similar to **Long-billed Corella** *C. tenuirostris*, which has very long bill and broad pink stripe across throat, and **Western Corella** *C. pastinator*, which has longer bill and more red on face. **DISTRIBUTION** Widespread across mainland, although absent from driest parts of WA and SA. Expanding coastally in eastern and southern Australia, and now present in eastern Tas. **HABITS AND HABITAT** Very gregarious and noisy, at times forming enormous flocks when foraging and roosting. Often associates with Galahs (see opposite) and Sulphur-crested Cockatoos (see below), taking seeds, shoots and roots from the ground. Favours open plains, savannahs and farmland, and also found around cities and towns.

Sulphur-crested Cockatoo ▪ *Cacatua galerita* 48–55cm

DESCRIPTION Unmistakable Australian cockatoo. Entirely white with conspicuous erectile yellow crest. Also subtle yellow wash on underside of wings, and large dark beak. Sexes are alike. **DISTRIBUTION** Found across northern, eastern and south-eastern mainland and Tas. Introduced to south-west WA in the early 1900s. Occurs inland as far as central Australia and western NSW. **HABITS AND HABITAT** A flock of calling Sulphur-crested Cockatoos is one of the noisiest wildlife sounds in Australia, where large flocks of birds can wheel around screeching at the same time. Highly gregarious, they can be seen from pairs to flocks of up to hundreds. They forage mainly on the ground, seeking out seeds, bulbous roots and flowers. Found in timbered areas usually in proximity to water, but also utilizes areas such as parks and sporting ovals.

Cockatiel ■ *Nymphicus hollandicus* 29–32cm

DESCRIPTION Easily identified small parrot. Entire body grey except for white shoulders on upperwing, and male has yellow face with orange cheek-patch and long, fine slightly

upcurved crest. Markings much subtler in female. **DISTRIBUTION** Spread across mainland, except Cape York (Qld) and some wetter coastal areas of southern and south-eastern Australia. Absent from Tas. **HABITS AND HABITAT** Usually found in pairs or small parties; strong flier. Forages mainly on the ground, searching out seeds of grasses, shrubs and grain crops. Frequents water points late in the day for drinking, and favours savannah and open woodland and forests. In inland areas often found in close association with Budgerigars (see p. 77).

Rainbow Lorikeet ■ *Trichoglossus moluccanus* 30cm

DESCRIPTION Unmistakable. Head purplish-blue, which is separated from bright green upperparts by narrow yellow-green collar (orange in northern subspecies). Underparts

comprise varying degrees of orange breast and blue belly, and sexes are alike. **DISTRIBUTION** Two subspecies, one found coastally around northern Australia from Kimberleys (WA) to Cape York (Qld); the other from here down east coast to Eyre Peninsula (SA). Introduced to Perth in the 1960s. **HABITS AND HABITAT** Often forms loud, fast-moving flocks and is highly gregarious. Mostly forages on flowers of shrubs and trees, particularly flowering eucalypts (both remnant and planted). Highly aggressive species that has thrived on improved urban habitats with high-yielding native trees. Uses wide variety of treed habitats, including rainforests and woodland, and is a common sight in urban areas.

Musk Lorikeet ▪ *Glossopsitta concinna* 20–23cm

DESCRIPTION Medium-sized, mostly green-bodied lorikeet, with yellow patch at sides of breast. Markings about the head distinctive, with blue crown, red forehead and red band extending behind eyes to ear-coverts. Mantle has olive wash. **DISTRIBUTION** Occurs around south-east corner of mainland, from north coast of NSW to broader Adelaide region of SA, also in eastern Tas. **HABITS AND HABITAT** Very gregarious and aggressive species, found in pairs, small groups and flocks of up to hundreds of birds within canopy of flowering trees. Active and a common sight in suburban areas such as parks and street trees. Also utilizes tall, dry, open eucalypt forest and woodland.

Purple-crowned Lorikeet

▪ *Glossopsitta porphyrocephala* 17–19cm

DESCRIPTION One of the smaller lorikeets. Generally dark green above and pastel blue below, with bright red underwings in flight. Distinctive head pattern comprising dark purple crown, band of orange-red across forehead and prominent orange patch on ear-coverts. **DISTRIBUTION** Found across southern Australia from east Gippsland to south-west WA, with range broken in middle by expanse of Nullarbor Plain. Central Vic. and south-west WA the strongholds. **HABITS AND HABITAT** Highly gregarious species, usually found in small groups, but at times may congregate at flowering trees in large flocks of up to hundreds of birds. Typically of lorikeets, highly active, noisy and conspicuous, clambering around foliage of flowering eucalypts to forage on nectar in flowers. Favours open, dry eucalypt forest and woodland, but also common in planted trees in cities and towns.

Double-eyed Fig-Parrot ▪ *Cyclopsitta diophthalma* 13–15cm

DESCRIPTION Very small parrot with unmistakable plump body and short tail. Several subspecies; in all body wholly green apart from lemon wash to sides of breast and dark primary feathers in wing. Heads have various combinations of red, blue and yellow in crown and face, depending on subspecies. **DISTRIBUTION** Subspecies *marshalli* found around Lockart River on Cape York (Qld), *macleayana* from Cooktown to near Mackay (Qld) and *coxeni* around Qld-NSW border. **HABITS AND HABITAT** Utilizes upland and coastal rainforests, particularly those containing fig trees (*Ficus* spp.), whose fruits provide food for the species. Usually found in large, continuous patches of habitat, rarely venturing to partly cleared or fragmented areas. Subspecies *coxeni* is endangered and sightings are extremely rare.

Australian King-Parrot

▪ *Alisterus scapularis* 40–45cm

DESCRIPTION Stunning and easily identified parrot. Male has rich scarlet head and body, dark green back and tail, and dark green wings with pale green patch. Female like male, except for green head and neck. **DISTRIBUTION** Found along east coast of mainland from around Cooktown in Qld to Otway Ranges in Vic. **HABITS AND HABITAT** Usually occurs in pairs or small flocks. Can become quite tame around humans. Notorious for eating fruits in orchards and gardens, as well as some commercial crops. Also able to use a variety of natural habitats, including dry eucalypt forest and woodland, and rainforests (both temperate and tropical).

Superb Parrot ■ *Polytelis swainsonii* 40cm

DESCRIPTION Large, long-tailed green parrot. Male bright green above and below, with dark primary feathers in wing. Facial markings distinctive, with bright yellow forehead, crown and throat bordered below by red line across throat. Male has red bill. Female paler green all over, lacking yellow-and-red facial markings of male. **DISTRIBUTION** Confined to inland slopes and adjacent plains of NSW and northern Vic., with Riverina of NSW and Vic. the stronghold. **HABITS AND HABITAT** Found from pairs to small parties; very swift flier. Exhibits some localized seasonal movements, being seen in Vic. and along Murray River mainly during summer breeding season. Forages on the ground on seeds, fruits and spilt grain, and also in trees. Mostly inhabits box-gum woodland on slopes, and River Red Gum forests and nearby woodland in Riverina.

Regent Parrot
■ *Polytelis anthopeplus* 37–42cm

DESCRIPTION Unmistakable bird that is mostly yellow with blue-black wings, bright yellow shoulder-patch and pale scarlet bar across wing. Female and young birds paler yellow than male, almost appearing greenish, and yellow and red in wings is less prominent. Both sexes have bright salmon-pink bill. **DISTRIBUTION** Two areas of distribution on mainland – in east found along lower Murray River and associated mallee of NSW, Vic. and SA; in west found broadly across south-west WA, including wheat belt area. **HABITS AND HABITAT** Highly gregarious; found in small parties to moderate-sized flocks. Mainly forages on the ground on seeds and vegetation, but also feeds in trees. Favours eucalypt woodland and forests, and also visits farmland. Known to visit orchards, which brings it into conflict with some farmers.

Crimson Rosella

■ *Platycercus elegans* 35–38cm

DESCRIPTION Interesting species with three main races, known as Crimson, Yellow and Adelaide Rosellas. Typical Crimson-type male has mostly crimson red plumage, with bright blue cheeks, blue shoulder-patches and blue tail. Female is duller. Juveniles have green bodies. Yellow Rosella has a yellow body, and Adelaide Rosella an orange-yellow body. **DISTRIBUTION** Typical Crimson Rosella found around south-east mainland from Kangaroo Island in SA, to north-east Qld between Atherton Tableland and Mackay region. **HABITS AND HABITAT** Occurs in pairs, family parties and small flocks, feeding in trees and shrubs on the ground. Usually associated with taller, wetter eucalypt forests, but also utilizes alpine woodland and farmland. Often found in parks and gardens, and can be quite tame and confiding.

Eastern Rosella ■ *Platycercus eximius* 28–33cm

DESCRIPTION Distinctive white-cheeked rosella with red head, neck and breast. Yellow back, yellow-and-green underparts, and red undertail-coverts. Dark wings with bright blue

shoulders. Female like male, but plumage generally duller. **DISTRIBUTION** Distributed around south-east of mainland from north of Brisbane (Qld) to south-east SA, and inland to Hay in NSW; also in Tas. **HABITS AND HABITAT** Gregarious; can be seen singly, but generally found in pairs or small flocks. Often feeds on the ground, but is quite furtive and hard to approach. Utilizes open eucalypt woodland and grassland, and is a common sight on farmland. Also found in urban areas like parks, gardens and golf courses.

Pale-headed Rosella ■ *Platycercus adscitus* 28–34cm

DESCRIPTION Distinctive appearance. Head very pale yellow with large white cheek-patch bordered with blue below, yellow back and pale blue underside from throat to vent. Dark wings and tail, and red undertail-coverts. **DISTRIBUTION** Found along east coast of Qld from tip of Cape York Peninsula down into northern edge of NSW; inland as far as Longreach in Qld. **HABITS AND HABITAT** Usually found in pairs or small parties, and often in association with Eastern Rosellas (see opposite). Arboreal and terrestrial; usually wary when foraging on the ground, but allows closer approach when feeding in trees. Favours savannah woodland, lightly timbered grassland and farmland. Quite noisy, active and conspicuous; mainly eat seeds and fruits.

Australian Ringneck

■ *Barnardius zonarius* 28–44cm; male larger than female

DESCRIPTION Several races across Australia, all with largely green bodies and tail, dark wings with blue stripe, and yellow half-collar around nape. Typical eastern form has dark blue-green back and red stripe over bill; typical western form has dark head and yellow belly. **DISTRIBUTION** Eastern form found around inland Qld, NSW, north-west Vic. and eastern SA; western form across southern WA, inland NT, and inland and south-west SA. **HABITS AND HABITAT** With a wide-ranging distribution, has adapted to variety of habitats, including woodland, timbered watercourses and shrubland. Gregarious; found in pairs and small flocks. Mainly terrestrial foraging but also arboreal, and can allow close approach when feeding in trees and shrubs. Generally inconspicuous, although can be quite vocal when in flocks.

Swift Parrot ■ *Lathamus discolor* 24–25cm

DESCRIPTION Largely bright green plumage with dark blue crown-patch, and prominent red face, chin and throat, narrowly bordered with yellow. Red shoulder and blue patch in

wing, and long, pointed maroon tail. Sexes differ slightly, with female being slightly duller than male. **DISTRIBUTION** Breeds in southern and eastern Tasmanian forests in spring and summer, crossing Bass Strait to overwinter across south-east mainland, anywhere from western Vic. to Brisbane region of Qld; inland as far as Riverina of NSW. **HABITS AND HABITAT** Specialized parrot that forages on nectar and psyllids, a leaf-sucking insect. Highly gregarious and usually noisy when feeding; occasionally found in communal roosts of hundreds of birds. Occupies temperate rainforests in breeding range, utilizes variety of temperate woodland and forests on mainland, and can take advantage of heavily flowering street trees.

Red-rumped Parrot ■ *Psephotus haematonotus* 24–30cm

DESCRIPTION Male generally bright green, with blue-green head, distinctive red rump, olive-green back, and yellow shoulders and belly. Female duller olive-green and lacks red rump of male. **DISTRIBUTION** Found across south-eastern mainland, including all of Vic., NSW (except far north coast), southern Qld and eastern SA. **HABITS AND HABITAT** Delightful and cheery little parrot that is very gregarious except when breeding. Often seen perched in trees or foraging on the ground. Favours open grassland or lightly timbered woodland and plains, as well as watercourses and farmland. Common sight in open, grassy spaces in towns and cities, including golf courses, sporting ovals and parks. Quite confiding and will allow close approach before flying away, giving sweet tinkling calls.

LEFT Male; RIGHT Female

Hooded Parrot ■ *Psephotus dissimilis* 26–28cm

DESCRIPTION Male has black forehead, cap and lores, olive-green back and turquoise-blue underside. Wings have golden-yellow shoulder-patch, and undertail-coverts are salmon-pink. Female has pale green back, and lacks black cap and yellow wing-patch of male. **DISTRIBUTION** Confined to Top End of NT, mainly near Pine Creek and Katherine-Mataranka; also across to west of Arnhem Land plateau. **HABITS AND HABITAT** One of three Australian parrots to nest in termite mounds, the others being the **Golden-shouldered Parrot** *P. chrysopterygius* of Cape York (Qld), and the now-extinct **Paradise Parrot** *P. pulcherrimus*. Usually in pairs or small flocks; quiet and confiding. When disturbed will fly to trees and wait for danger to pass before returning to feed on the ground. Found in monsoonal tropical open eucalypt woodland and lightly timbered grassland.

Male

Budgerigar ■ *Melopsittacus undulatus* 18cm

DESCRIPTION Unmistakable appearance, with largely green underside and yellow-and-black barred back and wings. Yellow head with black barring. Male has blue cere, which is pale brown in female. **DISTRIBUTION** Distributed across all mainland states and territories, although usually confined to inland and arid areas. Moves to coastal east and southern areas in drought, and further north in good seasons. **HABITS AND HABITAT** One of Australia's best-known parrots due to popularity as household pet. Irruptive species of the inland capable of breeding in massive numbers in good years, and can be found in tight-circling flocks of thousands of birds. Highly dependent on water within favoured habitat of grassland and rangeland.

Orange-bellied Parrot

■ *Neophema chrysogaster* 22–25cm

DESCRIPTION Bright grass-green parrot with bright yellow underside around rich orange central belly-patch. Rich blue forehead-band between eyes, and blue shoulder in wing. Female is slightly duller than male. **DISTRIBUTION** Breeds in remote south-west region of Tas. during summer, before migrating to mainland to overwinter in coastal habitat between Port Phillip Bay in Vic. and Coorong in SA. **HABITS AND HABITAT** Considered a bird of coastal saltmarsh habitat, it has undergone significant declines due to habitat loss and is now Critically Endangered and subject to an intensive recovery effort. Birds breed in hollows in eucalypts among buttongrass and coastal plains in Tas. Quiet and unobtrusive, they are quite wary of disturbance. Easily confused with other *Neophema* parrots by inexperienced observers.

Turquoise Parrot

■ *Neophema pulchella* 20–22cm

DESCRIPTION Stunning little parrot. Male grass-green on back, tail and parts of wing, and has vivid blue face and forehead. Underside bright yellow from throat to undertail, and wings have broad leading edge of vivid blue and red shoulder-patch. Female greener than male overall and lacks red shoulder. **DISTRIBUTION** Found on inland and coastal slopes and lowlands of Great Dividing Range of south-east Australia, from southern Qld to north-east Vic. **HABITS AND HABITAT** Usually found in pairs or small parties, but in non-breeding season can gather in flocks of up to 50 birds. Generally forages unobtrusively for seeds on the ground, and quickly flies to cover when disturbed. Declined in mid-1900s; decline appears to have been arrested due to concerted conservation efforts in some regions, such as north-east Vic.

Eastern Ground Parrot ■ *Pezoporus wallicus* 30cm

DESCRIPTION Long, slender grass-green parrot with yellow eye-ring and red forehead. Back and wings mottled black and yellow; underside barred black and yellow. Black streaking on crown.
DISTRIBUTION Found around Gympie and Fraser Island in Qld, along east coast of NSW and southern Vic., and across south-west Tas., where it is common
HABITS AND HABITAT Recently separated from similar **Western Ground Parrot** *P. flaviventris.* Favours coastal swamps and heathland, and nearby grassland. Extremely cryptic and furtive, and difficult to see well. Presence usually given away by loud, ascending calls at dawn and dusk. Terrestrial species, feeding on seeding sedges and rushes.

Eastern Koel ■ *Eudynamys orientalis* 41cm

DESCRIPTION Male wholly glossy black with red eyes and pale bill. Female has black head and throat-stripe, fawn underside with black barring, and grey-brown back with pale spots. **DISTRIBUTION** Found around Australia in spring and summer from Kimberley region of WA, across Top End and down east coast to around Canberra (ACT). Sighted with increasing frequency in southern NSW and Vic. in recent years. **HABITS AND HABITAT** Highly specialized egg parasite, using birds such as friarbirds, wattlebirds, Olive-backed Orioles (see p. 129) and figbirds as hosts. Extremely loud call that can start well before sunrise. Usually found singly or in pairs, and inhabits a range of habitats including tropical monsoon forest, wet eucalypt forest and open woodland. Main determinant of presence is fruiting trees.

Channel-billed Cuckoo

■ *Scythrops novaehollandiae* 56–70cm

DESCRIPTION Very large and unmistakable cuckoo. Head and breast light grey, barred slightly underneath. Back and wings dark grey with broad black barring. Red eye and eye-ring, and enormous curved, horn-coloured bill. **DISTRIBUTION** Similar to Eastern Koel (see p. 79); around northern and eastern mainland, south to eastern Vic. **HABITS AND HABITAT** Summer breeding migrant. Hosts include large birds like Pied Currawong, Australian Magpie and Torresian Crow (see pp. 131 and 134). Appears in pairs or small flocks; often cryptic with presence given away by calls. Very vocal when flying, at times wheeling about calling repeatedly. Forages in rainforests and open woodland on fruiting trees like figs (*Ficus* spp.). Also visits farmland and orchards to eat fruits, and known to take large insects like locusts, and occasionally eggs and nestlings of birds.

Black-eared Cuckoo

■ *Chrysococcyx osculans* 19cm

DESCRIPTION Brown-grey upperparts with pale grey rump. Creamy-buff underparts and grey-brown tail with prominent narrow white tip. Face creamy-white, with obvious black eye-stripe extending from beak to cheek. Sexes are alike. **DISTRIBUTION** Found widely across arid and semi-arid areas of Australia; also into Top End and drier coastal areas of southern Australia. Not found in waterless deserts. **HABITS AND HABITAT** Mostly seen singly, but also in pairs and small parties. Forages mainly on the ground, but also in foliage of trees and shrubs. Presence often given away by plaintive descending call. Host birds for this nest parasite include Speckled Warbler and Redthroat (see p. 100), scrubwrens and thornbills. Favours dry woodland and shrubland, mallee, mulga and lignum habitats.

Shining Bronze-Cuckoo ■ *Chrysococcyx lucidus* 13–18cm

DESCRIPTION Superficially similar to other bronze-cuckoos, but this species has a maroon-bronze head and iridescent green upperparts. Underparts white with neat dark barring that generally meets in middle (contra **Horsfield's Bronze-Cuckoo** *C. basalis*). Face white with fine dark mottling. Sexes are alike. **DISTRIBUTION** Found along eastern Australia from Cape York to Tas., around to Eyre Peninsula (SA), and also in south-west WA. Present October–April. **HABITS AND HABITAT** Migratory cuckoo that overwinters in Papua New Guinea and Indonesia, arriving in spring for the austral summer. Seen singly or in twos, and rarely in small groups. Normally inconspicuous when among foliage, but obvious in breeding season when males call from high perches. Uses wide array of wooded habitats, from lightly timbered woodland to rainforests. Eggs laid in nests of small songbirds like thornbills, fairy-wrens, flycatchers and honeyeaters.

Pallid Cuckoo ■ *Cacomantis pallidus* 31cm

DESCRIPTION Large cuckoo with slightly decurved bill, dark grey above and light grey below. Dark tail has notched white markings along margin, and dark face highlights yellow eye-ring. **DISTRIBUTION** Found right across mainland and Tas., arriving in southern areas in spring. Overwinters in central Australia or Papua New Guinea. **HABITS AND HABITAT** On arrival in breeding areas in spring birds are extremely vocal, with piping call carrying far. Favours woodland and scrubland habitats, but also found on farmland, in pastoral country and around cities and towns in parks and golf courses. Lays eggs in open, cup-shaped nests such as those of honeyeaters, robins, woodswallows and whistlers.

Fan-tailed Cuckoo ■ *Cacomantis flabelliformis* 24–28cm

DESCRIPTION Similar to **Chestnut-breasted** and **Brush Cuckoos** (*C. castaneiventris* and *C. variolosus*), this species has mid slate-grey head, back and wings, and is pale rufous below

fading to white vent. Bold dark grey-and-white barred undertail and yellow eye-ring. Sexes differ slightly; female usually greyer below than male. **DISTRIBUTION** Found in eastern mainland states, south and south-east of SA and south-west WA; also in Tas. Migrates from Papua New Guinea and northern Australia, arriving in southern Australia in spring. **HABITS AND HABITAT** Usually seen singly, and less commonly in twos or small groups. Forages for insects mainly on the ground, but often encountered sitting quietly on branch, stump or other perch. Prefers well-timbered habitats with well-developed understorey or ground layer, in which it is usually found within 10m of the ground. Also uses parks and gardens. Usually lays single egg in domed nests like those of fairy-wrens, thornbills and scrubwrens.

Powerful Owl
■ *Ninox strenua* 50–60cm

DESCRIPTION Largest Australian owl. Adults dark grey-brown above, flecked white, and whitish below with broad grey-brown mottling and 'V'-barring. Large yellow eyes and feet. Juveniles have white head, white underside with some brown streaking, and brown cap and face. **DISTRIBUTION** Restricted to suitable habitats along south-east corner of mainland, from south-east Qld to south-east SA. **HABITS AND HABITAT** Seen singly, in pairs and in family groups, the birds are an impressive sight. Often found during the day roosting on tree limbs or in dense mid-storey, standing on prey caught the previous night, such as possums and gliders. Breeding pairs have large home territories within favoured habitat of tall wet forests and gullies. Also found in woodland, coastal forests, and some capital cities and towns.

Barking Owl ■ *Ninox connivens* 39–44cm

DESCRIPTION Medium-sized owl with solid appearance. Dark grey head and back, and grey wings with whitish spots. Whitish underside with broad grey-brown streaking. Large, vivid yellow eyes. Smaller northern race more rufous. **DISTRIBUTION** Found in suitable habitat throughout eastern mainland states, northern NT and WA, and south-west WA. Most common in Qld and Kimberley (WA). **HABITS AND HABITAT** Bird of open and dry habitats dominated by eucalypts, particularly margins of wetlands, but also occurs in paperbark woodland and some scrubland. Opportunistically preys on birds and mammals during breeding season, and takes more insects during non-breeding season. Has declined substantially in southern part of range, but is still common in northern part. Main call like that of barking dog, a repeated *whoof whoof*.

Southern Boobook

■ *Ninox boobook* 30–35cm

DESCRIPTION Smallest hawk-owl on mainland; several races, which vary in size and plumage. Overall appearance is of small brown owl with white spotting on wings, and broad brown and white streaking, mottling or barring on underside. Facial disk rimmed with white giving 'goggled' appearance; eyes pale yellow. **DISTRIBUTION** Various races found across whole of Australia, except driest waterless deserts. **HABITS AND HABITAT** Found singly, in pairs and in family groups. Has adapted to occur in almost any wooded habitat that contains suitable hollows for nesting and roosting. Highly insectivorous at times, but usually takes birds and mammals as prey. Often allows close approach, even when roosting in dense vegetation during the day.

Greater Sooty Owl

■ *Tyto tenebricosa* 33–43cm; female larger than male

DESCRIPTION Solid-looking, sooty-black owl with large, dark grey facial disk bordered with black. Upperparts darker and contain small whitish spots, while underside has whitish spots or 'V' markings. Large dark eyes and dark legs. Smaller northern subspecies variously considered a distinct species, and is slightly paler than southern birds. **DISTRIBUTION** Found in suitable habitat along east coast from north of Brisbane (Qld) to north-east of Melbourne (Vic.). **HABITS AND HABITAT** Bird of tall, wet forests in sheltered gullies containing dense understorey. Very secretive; presence best found by listening for the unusual call – a loud, descending and wavering whistle akin to a falling bomb. Prey includes a range of arboreal and terrestrial mammals, and in some areas feeding sites are littered with thousands of bones. Occasionally takes birds and reptiles.

Masked Owl ■ *Tyto novaehollandiae*
33–50cm; female larger than male

DESCRIPTION Most variable of Australian owls in size and plumage, but typical form is rufous-brown above and pale below, with large, dark-rimmed facial disk typical of *Tyto* owls, and fully feathered legs. Upperparts have light spotting and some dark barring; underparts have faint dark spotting on flanks. Tasmanian birds largest and darkest (including dark face); northern birds smallest and palest. **DISTRIBUTION** Coastal mainland Australia, although patchily distributed. Widespread but threatened in NSW, Vic. and Tas. **HABITS AND HABITAT** Usually seen singly or in pairs, but at times occurs in family group with dependent young. Favours eucalypt forest and woodland, and adjacent open country, particularly at ecotone between the two. Strictly nocturnal, hunting by perch-and-pounce method from branches, stumps and fence posts for main prey of small terrestrial mammals. Easily confused with **Barn Owl** *T. alba*.

Azure Kingfisher
■ *Ceyx azureus* 16–19cm

DESCRIPTION Small, unmistakable kingfisher. Rich royal blue above and orange-chestnut below with white streak below head. Throat paler than belly, black beak and bright orange-red legs. **DISTRIBUTION** Found coastally around north, east and south-east Australia, including Tas., where it is endangered. Can be a long way from coast if suitable watercourses are present. HADITS AND HABITAT Found singly or in pairs. Inhabits banks of fresh or tidal creeks where there are roots and branches overhanging the water. Also utilizes swamps, billabongs and lagoons. Perches and scans water before diving head first to snatch fish and crustaceans from below the surface. Nest excavated in creek bank and birds incubate 5–7 eggs.

Buff-breasted Paradise-Kingfisher
■ *Tanysiptera sylvia* 30–36cm, including long tail streamers

DESCRIPTION Small, brightly coloured kingfisher. Crown and wings deep royal blue; back, rump and tail streamers white. Large orange-red bill, and chestnut-rufous underside from chin to vent. Immatures lack tail streamers and have dark bill. **DISTRIBUTION** Summer breeding migrant from Papua New Guinea, restricted to coastal north-east Qld from tip of Cape York to near Townsville. **HABITS AND HABITAT** Builds nest in chamber inside active termite mound, and female lays 3–4 eggs. Mostly found in monsoon rainforest with open clearings, gullies or vine scrub. Usually seen singly or in pairs. Generally shy, but active at dawn and dusk and after rain. Catches prey mainly on the ground, but also in mid-canopy, with invertebrates, frogs and lizards being the main component of the diet.

Laughing Kookaburra
■ *Dacelo novaeguineae* 40–48cm

DESCRIPTION Iconic and unmistakable bird. Generally brown on back and wings, and off-white below. Conspicuous dark brown eye-stripe and crown, whitish eyebrow, and rufous tail with broad darker barring. **DISTRIBUTION** Found in eastern Australian states, including south-east SA and Tas.; introduced and well established in south-west WA. **HABITS AND HABITAT** Known by early settlers as the 'bushman's clock', the species' raucous and far-carrying calls are a common sound at dawn in the Australian bush. Usually occurs in pairs or family parties. For a large bird, can be inconspicuous when foraging by perching quietly for long periods in wait for prey. Inhabits wide variety of eucalypt forest and woodland. Roosts communally.

Red-backed Kingfisher
■ *Todiramphus pyrrhopygius* 23cm

DESCRIPTION Small tree kingfisher with white collar and underside, and turquoise blue-green back, wings and tail. Head streaked dark grey-green, prominent black eye-stripe and broad sandy-red rump. Sexes are similar, with female's plumage being slightly duller than male's. **DISTRIBUTION** Widely found across mainland, except wetter, denser areas of south-west and south-east, and centre of Great Victoria Desert. Demonstrates seasonal movements. **HABITS AND HABITAT** Seen in singles or pairs, perching on open dead trees, branches and overhead wires. Well adapted to arid habitats, and often found far from water. Favours prey like insects and small reptiles, but also known to take frogs, birds and mammals. Nest a burrow and chamber built into wall of sandbank or river bed.

Sacred Kingfisher
■ *Todiramphus sanctus* 20–23cm

DESCRIPTION Turquoise-blue cap and upperparts, buff spot behind bill, buff-white underparts and broad cream collar. Broad black eye-stripe extends from base of bill to nape. Sexes are similar, although female's plumage is generally duller than male's. DISTRIBUTION Resident in northern Australia, and spring–summer migrant in south. Occurs in most of Australia, but absent from deserts of central and southern WA/SA. HABITS AND HABITAT The common tree kingfisher of temperate woodland. Seen singly, in pairs and in family parties. Often sits quietly on branch, post or overhead wire, searching for prey on the ground. Found in various eucalypt woodlands, mangroves and paperbark forests, tall open eucalypt forest and melaleuca forest. Quite conspicuous during spring and summer, with individuals calling often.

Rainbow Bee-eater
■ *Merops ornatus* 22–25 cm, including tail streamers

DESCRIPTION The only bee-eater in Australia. Very colourful plumage; golden crown, orange-yellow chin and upper throat, and black band on lower throat. Broad black eye-stripe. Green upperparts, with primary feathers coppery and black tipped. Green breast and blue lower abdomen. Tail and tail streamers black, often with blue tinge. DISTRIBUTION Found across Australia except driest waterless deserts, although it is a seasonal migrant to southern Australian over spring–summer only. HABITS AND HABITAT Very gregarious; usually found in pairs or small flocks. Noisy and conspicuous, with calls heard even when some distance away. Often seen perching on fence posts or overhead wires, from where it makes dashing flights after flying insects. Found in open forest and woodland, shrubland and cleared areas, usually near water. Nest a burrow and chamber excavated in sandbank, riverside or similar.

Oriental Dollarbird
▪ *Eurystomus orientalis* 26–29cm

DESCRIPTION The only roller in Australia. Stocky dark bird with brown head, blue-green upperparts and pale blue belly. Deep blue flight feathers, and prominent blue-white spot in wing in flight. Broad red bill, tipped black. **DISTRIBUTION** Migratory, arriving in Australia from Papua New Guinea in September–October and departing in March–April. Found around north and east of mainland, south to eastern Vic. **HABITS AND HABITA**T Usually seen singly or in pairs. Conspicuous and noisy, and most active in the afternoons and around dusk. Perches prominently in tops of trees, and can sit for lengthy periods of time before sallying for insects. Nests in tree hollows, favouring open woodland and trees around watercourses and wetlands.

Noisy Pitta
▪ *Pitta versicolor* 19–21cm

DESCRIPTION Head black with chestnut sides to crown; breast and flanks dirty yellow. Green back and wings with bright blue shoulder-patch. Black belly and red or pink undertail-coverts. **DISTRIBUTION** Coastal eastern Australia, from Cape York (Qld) to lower Hunter Valley (NSW); occasionally further south to Sydney and beyond. **HABITS AND HABITAT** The most widespread Australian pitta. Seen singly or in pairs. Quite shy and furtive, making close approach difficult. Usually seen foraging on the forest floor, turning leaf litter in search of insects. Also uses wood or stones to break snail shells. Favours tropical and subtropical rainforests, but also occurs in vine forest and wet eucalypt forest.

Superb Lyrebird ▪ *Menura novaehollandiae* 78–103cm, including 54–71cm-long tail

DESCRIPTION Large, ground-dwelling bird. Overall grey-brown appearance, with darker brown wings and chestnut throat. Both sexes have long tails, but male's is elaborate with beautifully patterned outer feathers framing 12 filamentous central feathers. **DISTRIBUTION** From Dandenong Ranges of Vic., north along Great Dividing Range to southern Qld. Introduced and established around Mt Field NP in Tas. **HABITS AND HABITAT** Male holds elaborate tail erect and forwards above head while displaying, at the same time performing some of the best mimicry in the avian world. Calls of a range of forest birds are intertwined with his own. Lives in temperate and subtropical rainforests, wet forests, woodland, fern gullies and adjacent gardens.

Noisy Scrub-bird
▪ *Atrichornis clamosus* 19–23cm

DESCRIPTION Dark brown above, with pale chest and rufous underside. Body finely barred. Short, pointed beak, and throat black with white lines down each side. Female has duller plumage than male and no black in throat. **DISTRIBUTION** Extremely range restricted; from Two Peoples Bay to Mt Manypeaks near Albany, south-west WA. **HABITS AND HABITAT** Incredibly cryptic bird that is heard but rarely seen, despite males having one of the loudest and most penetrating calls in Australia. Only seen singly, and usually poorly through sticks and vegetation as bird remains concealed. Occasionally darts across tracks or small clearings, running or hopping with a cocked tail. Found in dense undergrowth of wetter parts of coastal forests and heathland.

White-throated Treecreeper
■ *Cormobates leucophaea* 14–17cm

DESCRIPTION Adult plumage distinctive, being dark brown above and on sides of head, with distinctive white throat and chest, and white streaks on dark greyish flanks, edged black. Undertail-coverts barred. Female has small, rufous-orange spot on cheeks. **DISTRIBUTION** East and south-east Australia and islands, but absent from Bass Strait and Tas. **HABITS AND HABITAT** Usually seen singly or in pairs. Actively forages on trunks and larger branches of trees, creeping up and wheeling around as it searches for insects. Calls conspicuous and far carrying, and repeated often. Found in eucalypt forest, woodland and timbered river areas, and nests in tree hollows.

Brown Treecreeper
■ *Climacteris picumnus* 14–18cm

DESCRIPTION Easily identified treecreeper, with grey-brown upperparts and brownish underparts strongly streaked black and buff. Distinct dark streaking on ear-coverts and pale white eyebrow. Male has small patch of blackish feathers in throat, which are rufous in female. **DISTRIBUTION** Across eastern mainland in suitable habitat from Cape York to Vic., inland to western Qld, NSW and eastern SA. **HABITS AND HABITAT** Active, noisy and conspicuous. Resonating *plink* calls carry far and are made regularly. Gregarious; usually occurs in pairs or small groups. Hops across the ground and over fallen timber in search of food, as well as spiralling up tree trunks. Found in dry open forests and woodland, particularly those dominated by eucalypts and with a lot of fallen timber. Nests in tree hollows.

Rufous Treecreeper
▪ *Climacteris rufa* 16–18cm

DESCRIPTION The most distinctive treecreeper. Grey-brown above, reddish face, grey sides of neck and rich rufous underparts with faint creamy streaking. Male has patch of finely black-streaked feathers in centre of breast, which are buff in female. **DISTRIBUTION** Found from south-west WA across to Eyre Peninsula of SA, except for break in distribution on Nullarbor Plain, where there is no suitable habitat. **HABITS AND HABITAT** Usually seen in pairs or trios. Behaviourally very similar to Brown Treecreeper (see opposite). Forages on the ground, in fallen timber and leaf litter, but also wheels up and around trunks and branches of trees. Found in karri-jarrah forest and woodland, timbered watercourses, salmon gum and wandoo woodland further inland, and mallee-woodland in SA.

Green Catbird ▪ *Ailuroedus crassirostris* 31–32cm

DESCRIPTION Large greenish rainforest bird, with diagnostic call akin to child crying or cat meowing loudly (hence 'catbird'). Olive-green head and underside with fine, pale

white streaks, partial white ring around red eye, and sturdy whitish bill. Back, wings and tail plainer mossy-green, with white on tips of wing-coverts, secondaries and tail. **DISTRIBUTION** Endemic to central-eastern Australia, on and coastal of Great Dividing Range between Gladstone region of Qld and Narooma in NSW. **HABITS AND HABITAT** Usually seen singly or in pairs, with presence first given away by diagnostic call. Mainly found in subtropical rainforests; also in adjacent habitats and ecotones, and wet eucalypt forest. Occasionally occurs in gardens and orchards. Predominantly frugivorous, but also take insects, seeds, frogs and birds. Unlike other bowerbirds, does not make a bower.

Regent Bowerbird

■ *Sericulus chrysocephalus* 25–30cm

DESCRIPTION Adult male unmistakable. Wholly glossy black except for vivid yellow-orange crown, nape, collar and wing-patch. Yellow eye and bill. Female duller olive-brown with netting pattern on underside and white-centred feathers on back below dark nape-patch. **DISTRIBUTION** On and coastal of Great Dividing Range between Mackay region (Qld) and Sydney basin (NSW). **HABITS AND HABITAT** Usually seen singly, in pairs or in small flocks; at times congregates in larger numbers in winter. Unless habituated to people around parks and gardens, males are typically shy. 'Avenue' bower smaller and less boldly adorned than those of other bowerbirds. Mainly rainforest bird, but also occurs in contiguous wet forests, farmland, parks and gardens. Predominantly frugivorous.

Satin Bowerbird ■ *Ptilonorhynchus violaceus* 28–35cm

DESCRIPTION Male wholly glossy blue-black with violet-blue eyes and bluish-white bill. Female and immatures green above with brown wings, scaly olive-green pattern on pale underside, and brown tail. **DISTRIBUTION** Two races: one in highlands of north-east

Qld, the other coastal in south-east mainland in suitable habitat from south-east Qld to Otway Ranges, Vic. **HABITS AND HABITAT** Seen singly, in pairs or in small family groups. Male makes elaborate 'avenue' bower that is decorated with all manner of blue items, such as feathers, bottle tops and clothes pegs. Male takes up to seven years to reach full adult plumage. Mainly a frugivore of rainforests and wet forests; also frequents parks, gardens and orchards. Generally quite vocal.

LEFT Male; RIGHT Female

Spotted Bowerbird ■ *Chlamydera maculata* 27–31cm

DESCRIPTION Head, neck and underside generally pale rufous-brown with darker fishnet patterning; back, wings and tail darker brown with pale buff tips to feathers. Distinctive, small erectile pink crest on nape above grey band on lower hindneck. **DISTRIBUTION** Found throughout arid and semi-arid eastern Australia, from north-central Qld to southern NSW. Formerly occurred in north-west Vic., but few reliable records in recent decades. **HABITS AND HABITAT** Usually solitary during display and breeding seasons, although many young brown birds can congregate around small, 'avenue'-type bowers. Bowers typically decorated with whitish, green or red objects, including bones, shells and glass or shiny items near habitation. Forages singly or in small groups, mainly on fruits, but also takes seeds and insects. Inhabits inland scrub, Brigalow and inland or riverine woodland; also occurs around homesteads.

Splendid Fairy-wren ■ *Malurus splendens* 11–14cm

DESCRIPTION Exquisite fairy-wren, with adult male in breeding condition almost wholly glossy violet-blue all over. Black line through eye to collar on rear of neck, and black breast-band. Lighter iridescent blue cheek-patch. Female grey above and creamy below with light blue tail; male in eclipse plumage similar. There are several races, which differ slightly. **DISTRIBUTION** Description above is for race in south-west WA. Other races are in inland Qld, NSW and Vic. across to WA, with several breaks in range driving their differentiation. **HABITS AND HABITAT** Gregarious; typically found in pairs or small family groups. Carries tail cocked above back, and spends most of time hidden in vegetation – typically shyer than **Superb Fairy-wren** M. *cyaneus*. Utilizes range of habitats including dense shrubs in forests, woodland and areas around watercourses.

Red-backed Fairy-wren ■ *Malurus melanocephalus* 9–13cm

DESCRIPTION Smallest fairy-wren, with breeding male unmistakable; wholly jet-black with vivid crimson back. Female and eclipse male fawn-brown above, with pale fawn-white underside and no markings around eye (common in other fairy-wrens). **DISTRIBUTION** Found across northern and north-eastern mainland, from Broome (WA) to north-east NSW. **HABITS AND HABITAT** Found in tall grass in open woodland and swampy woodland; also scrubby understorey in rainforests, in gardens, and in tall grass on watercourses in inland areas. Gregarious; usually found in small flocks of mainly brown birds, and generally very vocal. Active, foraging by hopping on the ground or in shrubs like other fairy-wrens. Nest a typical oval-shaped dome; clutch typically 3–4 eggs, and earlier broods may help raise new young.

LEFT Male; RIGHT Female

Mallee Emu-wren
■ *Stipiturus mallee* 12–15cm

DESCRIPTION Small emu-wren with long, filamentous tail over 1.5 times body length. Rufous crown, rich lilac-blue face, throat and upper breast; back and wings grey-brown with dark centres to feathers; underparts pale buff-brown. Female duller than male, lacking blue throat. **DISTRIBUTION** Extremely range restricted; located in mallee eucalypt habitat in north-west Vic. and adjacent SA, although potentially extinct there due to recent extensive fires. **HABITS AND HABITAT** Endemic and highly endangered species, restricted to spinifex grassland on low dunes, with or without mallee eucalypt woodland. Secretive bird with faint, insect-like call; can be extremely wary. In fine weather often calls from top of bush or stick. Found singly, in pairs and in small groups, and often seen in association with, or near, Striated Grasswren (see opposite).

Striated Grasswren ■ *Amytornis striatus* 15–18cm

DESCRIPTION Medium-sized grasswren with slender body and long tail. Rusty red-brown above streaked white, orange eyebrow, black whisker mark in cheek, and white throat that blends with white-striped grey-brown underparts. **DISTRIBUTION** Patchily distributed through arid mainland, and common and widespread in inland WA; reaches central-west WA coast. **HABITS AND HABITAT** Usually seen in pairs, and often first detected by calls. Shy and secretive, making clear views difficult. Can approach in response to squeaking noises, clambering through spinifex to investigate. Mainly forages on the ground, searching leaf litter for invertebrates; found in spinifex grass associations of arid and semi-arid areas. Common in suitable habitat, but susceptible to wildfires and resultant loss of spinifex.

Black Grasswren ■ *Amytornis housei* 18–21cm

DESCRIPTION Second largest grasswren; male distinctive, largely black with silvery-white streaking, apart from rich chestnut upper wing, lower back and rump. Female similarly marked, but chestnut is lighter and continues across shoulders and underparts. **DISTRIBUTION** Extremely range restricted; found in suitable habitats in north-west Kimberley (WA). **HABITS AND HABITAT** Restricted to rough red-black sandstone escarpments, where it prefers to forage in shade of rock ledges, in areas of bare rock interspersed with tussock grasses like spinifex *Triodia*, or under shrubs. Gregarious; usually seen in pairs or small groups. Generally shy but male will sometimes pop up to sing on exposed ledges. Seldom flies – generally hops in typical grasswren style, or scurries rat-like across rocks.

Rufous Bristlebird
■ *Dasyornis broadbenti* 23–27cm

DESCRIPTION Largest bristlebird, with rich rufous-brown cap that extends over ears and nape, paler grey-brown upperparts and pale-scalloped greyish underparts. Red eyes. **DISTRIBUTION** Coastal western Vic. and south-east SA; race of south-west WA now considered extinct. **HABITS AND HABITAT** Usually found singly, rarely in pairs. Resonant call that sounds like a squeaky cartwheel, which is heard easily, but bird is much more difficult to see. Occurs in coastal scrub and heathland on dunes and cliff tops, and further inland along wet gullies in Vic. Rarely flies, and usually seen running along the ground from one patch of cover to another. Forages mainly on invertebrates, seeds and berries.

Pilotbird ■ *Pycnoptilus floccosus* 17–19cm
DESCRIPTION Small, plump bird. Rich buff-rufous with fine scalloping from forehead to breast; upperparts grey-brown and red eyes. Short, pointed black beak. Tail long

and broad, and wedge tipped. **DISTRIBUTION** Found in suitable habitat in far south-east mainland, and on coastal side of Great Dividing Range. **HABITS AND HABITAT** Common name reflects habit in some areas of following foraging Superb Lyrebirds (see p. 89). Usually seen singly or in twos, and rather tame and confiding. Terrestrial, at most a metre or two above the ground in shrubs or on logs, foraging for invertebrates in leaf litter and soil. Call loud and far carrying, and often easiest way to locate it. Rarely flies, and usually seen hopping along the ground flicking its tail up and down.

Rockwarbler ■ *Origma solitaria* 13–15cm

DESCRIPTION Small and scrub-wren like bird. Sooty olive-brown above, black tail, grey-white throat and rich rufous underparts. Forehead and rump washed rufous.

DISTRIBUTION Range restricted; found only in Hawkesbury sandstone and associated limestone regions of central and south-east NSW. **HABITS AND HABITAT** Sweet little bird of the sandstone escarpments, and the only bird endemic just to NSW. Hops around boulders and crevices, and into caves, at times appearing mouse like, in search of invertebrates. Nest an impressive hanging mass with side entrance; suspended from roofs or sloping walls of caves, overhanging rocks, road culverts and even inside garages. Most often seen singly or in pairs, but also in small family groups.

Yellow-throated Scrubwren ■ *Sericornis citreogularis* 12–15cm

DESCRIPTION Very distinctive, pretty bird. Male has black face-mask bordered above by yellow-white eyebrow, large yellow throat and olive-brown upperparts. Yellow throat

paler in female; face-mask olive-brown. Eyes red in both sexes; legs long and pinkish. **DISTRIBUTION** On and east of Great Dividing Range in two isolated populations; one from Cooktown to Paluma Ranges in Qld, the other from Kingaroy in Qld to south coast of NSW. **HABITS AND HABITAT** Endemic bird of rainforests and wet eucalypt forest. Seen singly or in pairs foraging either on the ground or in the understorey for invertebrates. Occasionally forages in association with Australian Logrunners (see p. 120) and **Chowchillas** *Orthonyx spaldingii*. Excellent mimic, generally chattering and making contact calls while foraging; song usually uttered from higher perch like log or shrub.

White-browed Scrubwren ■ *Sericornis frontalis* 11–15cm

DESCRIPTION Mostly dark olive-brown above, with pale cream or yellowish throat, and rufous flanks, belly and rump. Male has long white eyebrow and stripe below eye around

black face; female has olive-brown face. Pale yellow eye. Western race *maculata* has dark-spotted breast. **DISTRIBUTION** Three races, distributed around east, south and west coasts of mainland, and some Bass Strait islands. Absent from Tas. (where **Tasmanian Scrubwren** S. *humilis* is present). **HABITS AND HABITAT** Seen singly, in pairs or in small family parties. Usually found in dense undergrowth and lower levels of understorey shrubs, typically hopping briskly on the ground or on logs. Occupies diverse range of habitats, including rainforests, open forests, woodland and heaths; often in gullies and near watercourses where understorey is dense. Calls easily confused with Brown Thornbill's (see p. 104).

Scrubtit ■ *Acanthornis magnus* 11–12cm

DESCRIPTION Small, scrub-wren like bird, with rich rufous back, wings and tail, short white eyebrow, grey face and ear-coverts, pale eyes and white throat merging to rufous

underbelly. White tips to secondaries in wings. **DISTRIBUTION** Endemic to Tas., where it is widespread and fairly common everywhere except east coast. Also King Island in Bass Strait, where it is Critically Endangered. **HABITS AND HABITAT** Seen singly or in pairs; very shy and unobtrusive. Favoured habitat includes cool temperate rainforest dominated by Antarctic Beech *Nothofagus cunninghamii*, but also found in wet eucalypt forest, and prefers areas with dense understorey of ferns and shrubs. Forages in low and mid-understorey, hopping up trunks in search of invertebrates in method akin to foraging treecreepers.

Shy Heathwren
▪ *Hylacola cauta* 12–14cm

DESCRIPTION Small, terrestrial, scrub-wren like bird. Top of head, nape and back rufous-brown, reddish-chestnut rump, white eyebrow, whitish underside with broad black streaks, and conspicuous white chevrons in wing (contra **Chestnut-rumped Heathwren** C. *pyrrhopygia*). **DISTRIBUTION** Found patchily from central NSW and Vic., west through SA to southern WA. **HABITS AND HABITAT** Bird of mallee woodland and shrubland, cypress pine and mallee heath, usually seen singly or in pairs. Although shy it is the slightly bolder of the two heathwrens, so name is a slight misnomer. Considered good mimic of other birds' calls, and often sits on exposed perches to sing. Forages in leaf litter and understorey for invertebrates.

Striated Fieldwren
▪ *Calamanthus fuliginosus* 12–14cm

DESCRIPTION Forehead and face rufous-buff; rest of body olive-buff above and below, with coarse black striations throughout. Pale, striated throat and dark eyes. Short, pointed dark beak. **DISTRIBUTION** Coastal south-east Australia from Sydney basin (NSW) to Coorong (SA); also Tas. and Bass Strait islands. **HABITS AND HABITAT** Typically heard, then seen, sitting on exposed branch at top of shrub or fence post. Favours swampy or dry heathland and coastal scrubs, where seen mostly as singles or in pairs. When perched usually sits in hunched pose with tail cocked behind, swiped slightly from side to side or flicked vertically. Forages for invertebrates in low vegetation and leaf litter.

Redthroat ■ *Pyrrholaemus brunneus* 10–12cm

DESCRIPTION Bird of muted tones; head, neck and breast plain grey-brown, and wings and tail chestnut-grey. Underparts dull white, grading to buff. Pale lores and eye-ring around red eye, and male has small, rusty-reddish throat-patch. **DISTRIBUTION** Inland arid and semi-arid regions from Winton (Qld), south to north-west Vic., and west to WA coast between Moora and Pilbara. **HABITS AND HABITAT** One of the best songsters of inland Australia, its call has a canary-like quality and it is an excellent mimic of other songbirds. Favours acacia and chenopod shrublands and mallee, often along drainage lines. When not calling quite shy and unobtrusive, foraging in the understorey or on the ground. Domed nest with side entrance, in which clutch of 2–4 eggs is laid.

LEFT *Male;* RIGHT *Female*

Speckled Warbler ■ *Pyrrholaemus sagittatus* 11–13cm

DESCRIPTION Upperparts dark brown with buff, and underparts cream with bold black streaks. Tail dark with whitish tip. Male has black stripe over long white eyebrow;

female has chestnut stripe. **DISTRIBUTION** On and inland of Great Dividing Range, from Mackay region of Qld to Grampians of western Vic. **HABITS AND HABITAT** Gregarious; usually seen in pairs or small parties, and often mixed with other small insectivores like robins and thornbills. Often forages on the ground, among leaf litter or beneath shrubs or trees. Nest a grass bowl inside or under fallen timber. Found in dry eucalypt forest and woodland, *Callitris* woodland; in mulga and Brigalow in Qld.

Weebill ▪ *Smicrornis brevirostris* 8–10cm

DESCRIPTION Australia's smallest songbird; weighs just 6g. Plain, somewhat nondescript, grey-brown plumage, darker wings and buff-yellow underparts. Inland race has much more lemon underbelly. Short, stubby pale beak, whitish eye and creamy-white eyebrow. **DISTRIBUTION** Found around all of mainland, except treeless deserts, wettest forests and large agricultural areas. **HABITS AND HABITAT** Able to utilize wide array of woodland habitats, although tends to favour those dominated by eucalypts. These include box-ironbark woodland, mallee, Wandoo woodland, and woodland dominated by Coolabah in central Australia, and Darwin Stringybark further north. Gregarious, and almost always in association with thornbills and other canopy gleaners, searching foliage for leaf insects and lerp. Call distinctive and made often.

Western Gerygone ▪ *Gerygone fusca* 9–11cm

DESCRIPTION Plain grey-brown upperparts with darker wings; underparts white with grey wash on throat and breast. Faint white eyebrow over rich red eye, partial white eye-ring and black bill. Black tail has diagnostic white at tip and base. **DISTRIBUTION** Broadly inland of Great Dividing Range to central Qld, NSW and Vic.; then central Australia in southern NT-northern SA; south-west to central WA. **HABITS AND HABITAT** Most widespread gerygone, found in subtropical, temperate, arid and semi-arid areas in variety of habitats, including eucalypt woodland and forests, and acacia woodland, particularly where good understorey is present. Insectivorous, foraging through foliage for leaf insects, and probing bark. Unobtrusive; calls are best sign of its presence in a location.

Fairy Gerygone ■ *Gerygone palpebrosa* 10–12cm

DESCRIPTION There are two races. In northern race male has olive-green head, back and wings, and darker tail. Face blackish to throat, with white spot and white cheek-stripe; lemon underside. Female duller than male with no black face. Southern race paler – male as northern, but pale fawn underside, and no black face or white cheek. Female similar to

male. Races interbreed where they overlap. **DISTRIBUTION** Coastally in Qld; northern race from Cape York to Innisfail region, and southern race from Innisfail to near Brisbane. **HABITS AND HABITAT** Found singly or in pairs. Habitat includes edges of lowland rainforests, mangroves and dense riparian areas. Nest an impressive globular structure suspended from a branch, often overhanging water and near nest of paper wasps. Like other gerygones, generally unobtrusive and usually heard before being seen.

Chestnut-rumped Thornbill
■ *Acanthiza uropygialis* 9–11cm

DESCRIPTION Medium-sized thornbill. Upperparts grey-brown, and underparts creamy-whitish. Forehead finely scalloped in white, white eye and chestnut rump. Black tail has whitish tip. Black beak. **DISTRIBUTION** Inland southern and eastern Australia, south to coast of SA, and west to coast of WA. **HABITS AND HABITAT** Common sedentary species in inland Australia, usually seen in small flocks and often with other small passerines. Active, searching briskly on the ground and in low shrubs for insects. Conspicuous, regularly making contact calls. Found in dry woodland and shrubland, mainly mallee, mulga and other acacias; also chenopods shrubland. Nest small and domed, and placed inside hollow in tree, fence post or hollow stump.

Striated Thornbill ■ *Acanthiza lineata* 9–11cm

DESCRIPTION Medium-sized thornbill; greenish-brown back and tan coloured crown, with much white streaking on head and face. Underside pale yellow with darker, broader streaking on the throat and breast; dark eye. **DISTRIBUTION** Found on inland slopes and coastal side of Great Dividing Range, from near Bundaberg (Qld) around south-east mainland to Fleurieu Peninsula (SA). Absent Tas. **HABITS AND HABITAT** Gregarious, mostly in small flocks of 6 or more birds, and often in mixed-species flocks in autumn and winter with other thornbills, Weebill and fantails. Strictly arboreal, foraging by gleaning invertebrates from foliage or bark of mid- and upper-canopy of favoured eucalypt forest and woodland. Thin, insect-like contact calls often give away their presence well before they are seen.

Western Thornbill ■ *Acanthiza inornata* 8–10cm

DESCRIPTION Plainest thornbill. Olive-brown above, pale buff below with fine pale scalloping on dark forehead, and very pale buff rump. White eyes. **DISTRIBUTION** Endemic to south-west WA, from Moora region south-east to Stirling Ranges. **HABITS AND HABITAT** Found in pairs and small parties; joins mixed-species flocks with **Yellow-rumped Thornbills** *A. chrysorrhoa* and other small insectivores. Behaviour very similar to Buff-rumped Thornbill's (see above) of east. Unassuming and inconspicuous, but inquisitive enough to be called in by 'pishing' of observers. Mainly in eucalypt woodland and forests dominated by Wandoo, Jarrah and Marri. Nest domed and in shrubs, or hidden behind bark or in hollows; clutch size usually three eggs.

Brown Thornbill ■ *Acanthiza pusilla* 9–11cm

DESCRIPTION Olive-brown upperparts, with warm reddish-brown forehead. Rump reddish-brown, tail grey-brown with pale tip, and underparts off-white, streaked blackish on chin, throat and chest. Red eyes.

DISTRIBUTION On or adjacent to Great Dividing Range of eastern mainland from Proserpine (Qld) to Yorke Peninsula (SA); also Tas. and Bass Strait islands. **HABITS AND HABITAT** Common thornbill of eastern Australia. Can be seen singly, in pairs or in small family parties; restless, active and noisy. Mainly forages by gleaning insects from foliage, trunks and branches of shrubs and small trees. Found in wet and dry forests, woodland, shrubland, heathland and rainforests, as well as along watercourses, and parks and gardens in urban areas. Often found in association with White-browed Scrubwren (see p. 98).

Southern Whiteface ■ *Aphelocephala leucopsis* 10–12cm

DESCRIPTION Grey-brown above with blackish tail tipped with white, and creamy white below with grey wash across breast and on flanks. Face off-white. Sexes are alike.

DISTRIBUTION Found throughout southern Australia from west of Great Dividing Range to west coast of WA, although absent from south-west WA. **HABITS AND HABITAT** Gregarious; typically seen in pairs or small parties. Forages mainly on the ground, turning over leaf litter and other debris with bill. Inhabits woodland and shrubland with an understorey of grasses and shrubs; also grassland. Nest a domed structure built in hollows in stumps, trees or fence posts, or low in shrubs.

Spotted Pardalote ■ *Pardalotus punctatus* 8–10cm

DESCRIPTION Male has black head, wings and tail, which are covered with small white spots. White eyebrow, yellow throat, red rump and short black bill. Female is similar to male, but has yellowy spots on head and no yellow throat. DISTRIBUTION Found around eastern, southern and south-western regions of mainland; also in Tas. HABITS AND HABITAT A real jewel of the Australian bush; calls are part of 'soundtrack' of woodland birding. Usually occurs singly or in pairs, but in autumn–winter my join mixed-species flocks with thornbills, other pardalotes and Weebill (see p. 101). Active but inconspicuous, foraging high in tree canopy where it is readily revealed by its call. Favours eucalypt forest and woodland; also occurs in parks and gardens with well-established eucalypts. Nests in chamber at end of burrow in bank or similar.

LEFT Male; RIGHT Female

Forty-spotted Pardalote ■ *Pardalotus quadragintus* 9–11cm

DESCRIPTION Olive-green above, pale greyish below. Yellow-greenish face and undertail-coverts, dark tail, and dark wing-coverts and primary feathers with white spots on tips.

DISTRIBUTION Extremely range restricted; endemic to south-east Tas., where mostly found on Maria and Bruny Islands, and adjoining sections of main island. HABITS AND HABITAT Endangered species that has undergone serious declines in recent decades. Found almost exclusively in forest and woodland dominated by White Gum *Eucalyptus viminalis*. Like other pardalotes an extensive leaf gleaner, taking large amounts of lerp and psyllid insects. Nests in hollows in White Gums, which are thought to be scarce now and contributing to declines.

Striated Pardalote ▪ *Pardalotus striatus* 9–12cm

DESCRIPTION Variable, with as many as five recognized subspecies. Most common characteristics are black crown of head (sometimes streaked white), olive-brown back, dark

wings with white streak starting with yellow, red or orange spot, prominent white eyebrow that is yellow or orange over eye, and pale underside. **DISTRIBUTION** Widespread across mainland and Tas., in all regions except driest treeless deserts. **HABITS AND HABITAT** Found in wide variety of forests and woodland, often dominated by eucalypts, but also mulga, dry scrub and mangroves. Seen singly or in pairs during breeding season; during autumn and winter often in flocks. Mainly arboreal; gleans leaf foliage high in tree canopy. Nests in tree hollows or chambers at ends of burrows in riverbanks, road cuttings and cracks in building walls.

Western Spinebill ▪ *Acanthorhynchus superciliosus* 12–15cm

DESCRIPTION Unmistakable. Male has black crown and face-mask around white eyebrow. Grey back and wings, and pale belly. Broad chestnut collar bordered on lower

throat by white-and-black band. Distinctive long, downcurved bill. Female duller than male and lacks breast markings. **DISTRIBUTION** Endemic to south-west WA. **HABITS AND HABITAT** Most often seen singly or in twos, generally in loose association with other small honeyeaters. Favours heathland, particularly that with abundance of banksias, grevilleas and hakeas; also occurs in woodland dominated by Wandoo and Jarrah. Very active, moving rapidly between bushes on noisy wings to quickly feed on nectar with long, brush-tipped tongue. Also takes large amounts of insects on the wing, particularly when breeding.

White-lined Honeyeater ■ *Meliphaga albilineata* 17–20cm

DESCRIPTION Medium-sized, slender honeyeater; grey-brown above and cream below with buff mottling on breast and wash on flanks. Face and bill black, with curved white cheek-line below eye from yellow gape to rear of face. **DISTRIBUTION** Endemic to northern mainland, with disjunct populations in northern Kimberley (WA) and Arnhem Land (NT); thought by some to be separate species (Kimberley Honeyeater in WA). **HABITS AND HABITAT** Usually seen singly or in pairs, but congregates at times around flowering trees or water. Typically shy and hard to find; often heard but not seen as it forages in canopy of dense trees and shrubs. Inhabits cliffs and plateaus of escarpments, typically in gullies and gorges.

Bridled Honeyeater ■ *Bolemoreus frenatus* 20–22cm

DESCRIPTION Medium-sized honeyeater. Adults mostly grey-brown above and slightly paler below. Distinctive face markings; yellow base to bill, white gape to under eye where it joins small, pink fleshy patch, white 'bridle' above rear of eye and small yellow streak in dark ear-coverts. **DISTRIBUTION** Endemic to north-east Qld, from Cooktown area south to near Townsville. **HABITS AND HABITAT** Usually seen singly or in twos in highland tropical rainforest, wet eucalypt forest and wetter eucalypt woodland. Active and noisy; aggressive species and can become quite tame, coming to feeders. Generally forages in upper canopy on nectar and fruits, but also readily hawks insects on the wing.

Yellow-faced Honeyeater

■ *Caligavis chrysops* 15–18cm

DESCRIPTION Very pretty, almost wholly olive-brown honeyeater. Off-white with grey streaking on underparts, with distinctive downcurving yellow facial streak bordered inside black mask, and slight yellow mark behind eye. Slightly downcurved dark bill. **DISTRIBUTION** Found along and on plains of Great Dividing Range from Cooktown (Qld) through south-east Australia to Fleurieu Peninsula (SA). Partly migratory. **HABITS AND HABITAT** Birds seen in southern Australia are generally seasonal migrants, arriving in spring from northern parts of range, and departing in autumn. Some birds overwinter in south. Travels in flocks with **White-naped Honeyeater** *Melithreptus lunatus*, and often found in association with other similar-sized honeyeaters in eucalypt-dominated woodland and forests, heathland, parks and gardens. Usually seen in singles, pairs or small groups.

Singing Honeyeater ■ *Gavicalis virescens* 16–24cm

DESCRIPTION Medium-sized masked honeyeater. Upperparts plain grey-brown with faint trace of yellow in wings; underparts grey-white with dark grey-brown streaks. Distinctive

facial markings; dark stripe through eye curves down sides of neck, and bordered below eye by yellow streak and white ear-coverts. Sexes are alike. **DISTRIBUTION** Widespread and common. Found across mainland and drier islands; also coastal in central Qld, western Vic. to Melbourne, and southern and western coast. **HABITS AND HABITAT** Occurs singly, in pairs or in small groups. Pugnacious, often chasing other birds, including those larger than itself. Habitat used is mainly open shrubland and low woodland, especially those dominated by acacias; also coastal scrubs. Often seen in urban parks and gardens, and on farmland.

White-gaped Honeyeater
■ *Stomiopera unicolor* 17–22cm; male larger than female

DESCRIPTION Drab, medium-sized honeyeater. Almost uniformly dark grey-brown body, paler underneath with creamy mottling, and white semi-circular patch on and above gape. Robust, slightly curved black bill. **DISTRIBUTION** Found across tropical northern Australia and coastal islands, from Kimberley region (WA) to Proserpine (Qld); break in distribution in western Qld Gulf country. **HABITS AND HABITAT** Generally common and sedentary bird, usually found in pairs or small parties. Forages on nectar, fruits and insects in mainly riverine forests and woodland, and in other riparian vegetation like *Pandanus*, paperbarks and mangroves; also in towns and gardens. Active, noisy and aggressive; often seen in group chases and displays.

White-eared Honeyeater ■ *Nesoptilotis leucotis* 17–21cm

DESCRIPTION Distinctive; body olive-green on upperparts, yellow-green below. Crown of head grey; face and throat black with large white 'half-moon' patch over ear. **DISTRIBUTION** Widespread but patchily distributed from south-central Qld, through south-east mainland across Nullarbor into south-west WA wheat belt. **HABITS AND HABITAT** Found singly or in pairs; slightly nervous disposition, but will also inquisitively approach 'pishing' by observers. Found in range of habitats, including dry eucalypt forest and woodland, particularly mallee. Usually prefers habitat with well-developed shrubby understorey. Also in coastal scrub and heathland, and known to attack fruits in orchards. Aggressive to other species, particularly around concentrated food sources.

Yellow-tufted Honeyeater
■ *Lichenostomus melanops* 17–21cm

DESCRIPTION Several subspecies, all dark olive on back, wings and tail, and with underparts olive-yellow to yellow. Large, elongated black face-mask with distinctive pointed yellow ear-tufts, and yellow throat with dark central stripe. Race *cassidix* has small, erectile yellow crest on forehead. **DISTRIBUTION** On, inland and coastal of Great Dividing Range from southern Qld to western Vic. **HABITS AND HABITAT** Sedentary and gregarious; seen from pairs up to small groups of ten or more birds. Most races found in eucalypt forest and woodland, with dense understorey in coastal areas and open understorey inland of the ranges. Also Brigalow, mallee and cypress pine woodlands. Critically Endangered race *cassidix* found in tall, wet swamp forest of Mountain Swamp Gum east of Melbourne.

Yellow-plumed Honeyeater ■ *Ptilotula ornata* 14–18cm

DESCRIPTION Olive-green above, with slightly yellower head. Distinctive large, upswept yellow plume around lower face and rear of face. Underparts whitish with broad grey-brown streaks. **DISTRIBUTION** From central mallee of NSW, westwards through mallee of central and north-west Vic. to western SA, and south-west WA. **HABITS AND**

HABITAT Most often associated with mallee, and can survive well far from water. Gregarious at times, but mostly seen singly or in pairs. Very active forager in outer foliage and allows close approach; very aggressive with con-specifics and other species. Also found in coastal heaths, eucalypt woodland, River Red Gums on watercourses and coastal forests of south-west WA.

Fuscous Honeyeater
■ *Ptilotula fuscus* 14–17cm

DESCRIPTION Rather plain, medium-sized honeyeater. Olive-brown above, with paler grey-brown breast and buff off-white underparts. Small yellow neck plume, and yellow eye-ring, gape and base of bill. In breeding condition bill and eye-ring turn black. **DISTRIBUTION** Endemic to east and south-east mainland, from central Qld coast to western Vic.-eastern SA. **HABITS AND HABITAT** Usually seen in pairs or small flocks, and often in association with other honeyeaters like Yellow-tufted Honeyeater (see opposite), **White-naped Honeyeater** *Melithreptus lunatus* and Yellow-faced Honeyeater (see p. 108). Largely sedentary but capable of large movements in search of big flowering events. Active bird, busily foraging through flowers in canopy or hawking insects. Inhabits dry, open eucalypt forest and woodland with a shrubby understorey; sometimes occurs on farms and in gardens.

White-plumed Honeyeater
■ *Ptilotula penicillata* 15–18cm

DESCRIPTION Yellowish olive-grey above and pale brown-grey below, with yellowish head and face with distinctive white neck-plume below black line. Sexes are alike, but female is slightly smaller than male. **DISTRIBUTION** Widespread across east and south-east Australia, and in band through central Australia to WA coast. **HABITS AND HABITAT** Very common and highly territorial, gregarious species. Feeds in canopy and outer branches of trees, constantly moving from tree to tree with rapid darting movements. Found in open forests and woodland, often near wetlands and watercourses; inland tends to be tied to River Red Gums along rivers. Also in urban areas like parks and gardens, where it can dominate planted, heavily flowering native shrubs.

Noisy Miner ■ *Manorina melanocephala* 24–28cm

DESCRIPTION Large honeyeater, with dark grey-brown back and tail, and grey rump. Underside pale grey, with darker scalloped breast. Head has black saddle over crown and

whole face, bare yellow skin behind eye and yellow bill. Sexes are alike. **DISTRIBUTION** Coastal and inland eastern Australia from south-east Cape York (Qld) to Yorke Peninsula (SA). **HABITS AND HABITAT** Highly gregarious, living in communal groups that aggressively defend territories, and suppress diversity and abundance of smaller birds like whistlers, thornbills and other honeyeaters. Forages in foliage from canopy to low-shrub layer, and sometimes on the ground; takes nectar but also large amounts of lerp. Utilizes woodland and open forests, predominantly eucalypt dominated, especially with open or sparse understorey. Also occurs in urban areas, where it is common in parks and gardens.

Spiny-cheeked Honeyeater ■ *Acanthagenys rufogularis* 22–27cm

DESCRIPTION Boldly marked and slightly comical-looking, large honeyeater; olive-grey head, back and wings, pale grey rump and white-tipped, grey-brown tale. Dark-tipped

pink bill, with vivid pink gape running below eye to meet tuft of spiny feathers. Apricot-buff throat and creamy underbody with broad brown teardrop-shaped pattern. **DISTRIBUTION** Widespread on mainland, except tropical areas and parts of south-east Australia. **HABITS AND HABITAT** Usually found in singles to small family groups. Often active and conspicuous, but at times can be shy. Call unique and sounds 'unfinished', with many notes making it seem as though there is more to learn. Found predominantly in arid and semi-arid zone woodland dominated by acacias, but also with eucalypt-dominated habitat like mallee.

Regent Honeyeater ■ *Anthochaera phrygia* 20–24cm

DESCRIPTION Medium-large honeyeater with striking black-and-yellow plumage. Black head down to back and breast gives hooded appearance, with bare warty facial skin around eye. Back black with pale yellow edges to feathers; distinctive black-and-yellow scalloped breast fades to pale yellow belly. Striking vivid yellow patches in wings, tail and undertail. **DISTRIBUTION** Distributed through south-east Australia, mainly inland of Great Dividing Range but also coastal, from southern Qld to central Vic. **HABITS AND HABITAT** A rich patch nomad of eucalypt woodland and open forests, following flowering events of key eucalypt species like Mugga Ironbark and White Box around the landscape. Seen mostly in small groups, but can be found in flocks of up to 50 birds. Has, however, declined precipitously and is now Critically Endangered and subject to long-running recovery effort.

LEFT Male; RIGHT Female

Red Wattlebird

■ *Anthochaera carunculata* 33–37cm

DESCRIPTION Large honeyeater. Head has black crown, silvery-white face with red eye, and distinctive pendulous red wattle in cheek. Body and wings dark grey with silvery-white streaks, bright yellow patch on belly and long, dark grey tail. **DISTRIBUTION** South-east, southern and south-west mainland, from Qld-NSW border to near Shark Bay (WA). **HABITS AND HABITAT** Gregarious species, seen singly or in pairs when breeding, but sometimes in large non-breeding flocks. Very aggressive to other birds at times, and vocal and conspicuous. Forages in eucalypt woodland, forests and heathland, as well as mallee, but also common in towns and cities, where it takes advantage of planted native trees and shrubs.

White-fronted Chat ■ *Epthianura albifrons* 11–13cm

DESCRIPTION Distinctive tubby appearance. Male has white face, throat and belly, with black crown that extends to neck and forms band across breast. Silvery-grey back and wings, and black tail. Female more grey-brown than male, with less distinct breast-band.
DISTRIBUTION South-east, southern and south-west mainland, from Qld-NSW border to near Shark Bay (WA); inland in south-east to north-west NSW and adjoining SA. **HABITS AND HABITAT** Generally seen in small groups. Active and conspicuous, often foraging on bare or grassy ground. Favours saltmarsh and other damp areas with low vegetation, such as swampy farmland and roadside verges, but inland also uses saltbush plains and dune country around salt lakes and plains.

LEFT Male; RIGHT Female

Gibberbird
■ *Ashbyia lovensis* 11–14cm

DESCRIPTION Small, pretty ground bird. Sandy-fawn above with darker wings and tail, and pale lemon face, breast and underbody. Pale yellow eyes, and faint fawn stripe through eye. Dark bill. Female is paler than male.
DISTRIBUTION Inland eastern Australia, broadly around convergence of borders of NSW, Qld, SA and NT; most common around Lake Eyre basin. **HABITS AND HABITAT** Bird of gibber plains of central Australia, where mostly found on sparsely vegetated plains but also stony ridges. Usually seen in twos. Strictly terrestrial and forages by running after insects, or digging and turning over claypan. Tame and confiding, allowing close approach. Often seen in association with Australasian Pipit (see p. 151). Nests in deep cup built in depression on the ground.

Black Honeyeater ■ *Sugomel nigrum* 10–13cm

DESCRIPTION Male black on upperparts and tail, with black throat forming wedge into white underbody on breast. Dark eye and longish, fine, downcurved black bill. Female

mottled fawn-brown, darker above than below. **DISTRIBUTION** Arid inland Australia, west to WA coast. **HABITS AND HABITAT** Mostly seen singly, in pairs or in small, loose flocks. Moves widely to follow blossom and nectar flows. Often found in association with **Pied Honeyeater** *Certhionyx variegatus* and other species at flowering events. Behaviour reminiscent of spinebills, darting from bush to bush to feed on nectar or hawk insects. Found in arid and semi-arid woodland and mallee, particularly those areas with flowering emu-bush *Eremophila* species.

Scarlet Honeyeater ■ *Myzomela sanguinolenta* 9–11cm

DESCRIPTION Small honeyeater with long, downcurved dark bill. Male has rich crimson head, upperback, rump and breast, with brown-black wings and tail, and white belly.

Female almost wholly grey-brown, paler below with faint crimson wash on chin and throat. **DISTRIBUTION** Generally coastal eastern Australia in suitable habitat, from Cooktown (Qld) to Gippsland (Vic.), although may be expanding range as far as central Vic. **HABITS AND HABITAT** Found in singles or twos. In response to large flowering events can descend on areas in hundreds. Appears resident in north of range, migratory in south. Active and conspicuous in canopy of trees, where it forages in blossom or hawks insects. Favours eucalypt forest and woodland, particularly around wetlands, and occasionally rainforests.

Tawny-crowned Honeyeater ■ *Gliciphila melanops* 14–18cm

DESCRIPTION Medium-sized, slender and graceful honeyeater. Brown above and white below, with long, downcurved black bill. Cinnamon-buff crown, white eyebrow and dark

face-mask that continues down sides of throat reminiscent of saddle. Sexes are alike. **DISTRIBUTION** From north-east coast of NSW, through Vic. and SA where also in inland mallee areas, and south-west WA. Also coastal Tas. **HABITS AND HABITAT** Seen singly or in pairs. Shy and furtive, and usually detected by plaintive piping song. Forages in low shrubs for nectar and insects. Preferred habitat is heathland of temperate zone, but does occur in sand-plains. Also uses eucalypt woodland with dense heathy understorey.

New Holland Honeyeater
■ *Phylidonyris novaehollandiae* 16–20cm

DESCRIPTION Medium-sized honeyeater, mostly black above with faint white streaks on back, and white below with broader black streaking. Large yellow wing-panel, and yellow sides on tail. Head distinctive; small white eyebrow, tuft below eye and larger white tuft in ear-coverts. Sexes are alike. **DISTRIBUTION** In south-east and southern Australia, from Brisbane region (Qld) to north of Perth (WA); also in Tas. **HABITS AND HABITAT** Usually seen in small groups of up to ten birds; active and pugnacious, regularly chasing other honeyeaters from food plants. Flight fast and often erratic. Found in forests and woodland, especially those with dense shrub layer, and common in urban parks and gardens, where it exploits planted native trees and shrubs. Forages in all levels of remnant habitat.

Brown-headed Honeyeater ■ *Melithreptus brevirostris* 12–14cm

DESCRIPTION Small, plain honeyeater; olive-green wings, back and tail, and pale grey to buff below. Head brown with pale creamy stripe across nape and creamy yellow eye-ring.

Sexes are alike. **DISTRIBUTION** Broadly around south-east mainland, from central Qld to start of Nullarbor Plain, SA; also south-west WA. **HABITS AND HABITAT** Gregarious; usually seen in small flocks foraging high in tree canopy or flying overhead. Flight is acrobatic, and birds are always on the move. Noisy during flight and when foraging, uttering constant contact calls and song. Favours open, eucalypt-dominated forests and woodland, and often seen in urban parks and gardens. Forages on nectar in flowering trees and shrubs, and on insects.

Black-headed Honeyeater ■ *Melithreptus affinis* 14cm

DESCRIPTION Small 'hooded' honeyeater. Black head to throat, continues in slight saddle down sides of breast. Small, bluish-white wattle over top half of dark eye. Body olive-green above, white below.

Short dark bill. **DISTRIBUTION** Endemic to Tas. and Bass Strait islands; most common in northern and eastern Tas. **HABITS AND HABITAT** Gregarious; usually seen in small flocks. In autumn readily joins mixed-species flocks with **Strong-billed Honeyeater** M. *validirostris*, thornbills and pardalotes. Most foraging is in foliage of trees and limbs in canopy, but occasionally occurs lower down. Favoured habitat is dry sclerophyll forest dominated by eucalypts, and below 1,000m elevation. Nest a deep cup hanging in foliage or branches.

Blue-faced Honeyeater ■ *Entomyzon cyanotis* 28–32cm

DESCRIPTION Large, unmistakable honeyeater. Body olive-green above and white below, with black head, neck and bib, and striking blue skin around white eye. White line from

gape to breast encircles bib, and small white nape-crescent. Immature birds have olive-green facial skin.
DISTRIBUTION Around northern and eastern Australia, from Kimberley (WA) to south-east SA, in tropical, subtropical and temperate zones.
HABITS AND HABITAT Noisy and gregarious; usually seen in pairs or small flocks. Forages mainly in trees on insects on bark and limbs, among foliage (often acrobatically) and at flowers. Found in open forests and woodland, close to water, such as River Red Gum-lined streams. Also seen in orchards, farmland, and urban parks, gardens and golf courses. Generally confiding.

Noisy Friarbird
■ *Philemon corniculatus* 30–35cm

DESCRIPTION Large friarbird; bald head reminiscent of vulture's head, with knob on upper bill and faint white plume above red eye. Bib of silvery-white plumes on throat; upperbody grey-brown, paler below.
DISTRIBUTION East coast of mainland, from tip of Cape York (Qld) to eastern Vic., then west to SA along Murray River corridor.
HABITS AND HABITAT Gregarious; often seen in small to large flocks. Exhibits seasonal movements, although probably closely related to flowering events. Can descend en masse to favoured eucalypt woodland and forests in good blossom, providing deafening spectacle. Also uses heathland and coastal scrub. Active, incredibly noisy and conspicuous. Highly arboreal, foraging in outer foliage for nectar and insects. Nest a massive, deep cup hung in fork of end of branch.

Painted Honeyeater ■ *Grantiella picta* 14–15cm

DESCRIPTION Elegant medium-sized honeyeater. Male has black head, face, wings and tail, with bright yellow panels in wings and tail. Underparts white with some fine black streaks on flanks. Red eyes and deep pink bill. Female is duller than male and lacks streaking. **DISTRIBUTION** Eastern and northern Australia, from bottom of Gulf of Carpentaria through inland plains of Qld and NSW to western Vic. **HABITS AND HABITAT** Closely associated with fruiting mistletoe, but also takes nectar and insects. Usually seen singly or in pairs, less often in small flocks. Call delightful and far carrying, and often gives away bird's presence. Seasonal migrant, in southern Australia spring–summer only. Found in dry, open box-ironbark forest and woodland on inland plains of Great Dividing Range, but also in stands of acacia.

Grey-crowned Babbler ■ *Pomatostomus temporalis* 23–26cm

DESCRIPTION Largest babbler in Australia. Adults mostly dark brownish-grey, browner below, with distinctive head pattern of grey crown, broad white eyebrow, dark face and white throat. Long, downcurved black bill and yellow eyes; black tail tipped with white. **DISTRIBUTION** Across north-west, north, central and eastern Australia to Vic. **HABITS AND HABITAT** Gregarious and highly social bird, living in family groups. Very vocal, known colloquially in some areas as 'yahoo' bird for calls and animated behaviour. Extremely active, mainly terrestrial insectivore. Employs communal roost 'nests' – large, round, domed structures shared by whole family. Favours open forests and woodland, scrub and remnant farmland trees; at times occurs in towns and city gardens.

White-browed Babbler ■ *Pomatostomus superciliosus* 17–21cm

DESCRIPTION Smallest Australian babbler. Largely dark grey-brown to brown above with white throat and breast blending to brown belly. Fine white eyebrow, dark eye-stripe around

dark eye and downcurved black bill. Sexes are alike. **DISTRIBUTION** Found across southern Australia from west of Great Dividing Range in east to WA coast, generally south of Tropic of Capricorn. **HABITS AND HABITAT** Gregarious; often seen in groups of up to 15 birds. Active and noisy, foraging mostly on the ground among leaf litter and vegetation while constantly making contact calls. Inhabits dry sclerophyll woodland and open forests with a shrubby understorey; common in appropriate mulga woodland. Employs separate brood and communal roost nests.

Australian Logrunner ■ *Orthonyx temminckii* 18–20cm

DESCRIPTION Unmistakable, plump terrestrial bird. Largely rufous-brown upperparts with much black scalloping, and dark wings with grey fringes to coverts. Grey face-mask extends down sides of neck (where bordered black) and breast to flanks. Lower breast white and belly dark brown. Male has white throat and upper breast; rufous in female. **DISTRIBUTION** Coastal south-east Qld to southern NSW. **HABITS AND HABITAT** Most often encountered in pairs, and usually quiet and unobtrusive. Very vocal at dawn, with loud call carrying far. Works through leaf litter and fallen timber in search of invertebrates; allows close approach if observer is quiet and patient. Found mostly in subtropical and temperate rainforests and adjacent wet sclerophyll forest.

LEFT Female; RIGHT Male

Spotted Quail-thrush ■ *Cinclosoma punctatum* 25–29cm

DESCRIPTION Largest quail-thrush. Rufous-brown above with black scalloping; wings black with rufous-chestnut patch. Distinctive face markings on small head. Fine white eyebrow, dark face and throat, and white cheek in male; brown face and rufous cheek in female. Grey breast with clear demarcation to white underbelly and black-spotted flanks in male. Female more buff below. **DISTRIBUTION** On and in foothills of Great Dividing Range of south-east mainland from south-east Qld to south-east SA; east coast of Tas. **HABITS AND HABITAT** Found in dry, open sclerophyll forests and woodland, usually eucalypt dominated, on stony ridges and slopes. Predominantly terrestrial although occasionally in lower shrub level, where it searches for invertebrates. Found singly or in pairs. Usually quiet and unobtrusive, and most often located by listening for high-pitched *seeee* call. Furtive.

LEFT Male; RIGHT Female

Chestnut Quail-thrush ■ *Cinclosoma castanotum* 22–26cm

DESCRIPTION Medium-sized quail-thrush. Olive-brown above with deep chestnut rump, lower back and scapulars. White fringes to black wing-coverts. Underbody white with black from chin to low on breast bordered grey in male, all grey in female. Fine white eyebrow and thicker lower cheek-stripe. **DISTRIBUTION** Across southern and central mainland, from western Vic.-southern NSW to wheat belt of south-west WA. **HABITS AND HABITAT** Mostly seen singly, in pairs or in small family groups. Very unobtrusive and furtive like other quail-thrushes. Found in arid and semi-arid areas in low, shrubby undergrowth of mallee woodland, but also uses acacia shrubland and heathland. Feeds on arthropods and seeds, and occasionally on fruits.

Eastern Whipbird ■ *Psophodes olivaceus* 25–29cm

DESCRIPTION Distinctive appearance. Adults have black head, nape and breast; erectile crest on crown, large white patches on sides of throat and rich olive-green upperparts.

Underparts mottled whitish in olive-green, and fanned tail white tipped. Immatures much duller than adults. **DISTRIBUTION** Coastal eastern Australia, from near Cooktown (Qld), through NSW and Vic., to north-east of Melbourne. **HABITS AND HABITAT** Secretive bird whose antiphonal calls are heard more than birds are seen. Male gives loud and resonant 'whipcrack' calls, replied to immediately by female with quieter *choo choo*. Most often associated with thick understorey of dense wet forests, but also uses coastal shrubland and similar. Insectivorous; usually in pairs.

Chirruping Wedgebill ■ *Psophodes cristatus* 19–20cm

DESCRIPTION Medium-sized bird with distinctive, slightly recurved, pointed crest on top of head. Overall drab olive-grey plumage, paler below with darker streaks, and white-tipped black tail. Dark eyes and

bill. **DISTRIBUTION** Arid eastern inland Australia, in south-west Qld, north-west NSW and north-east SA. **HABITS AND HABITAT** For a long time considered the same species as **Chiming Wedgebill** *P. occidentalis* of central and western inland Australia, this species is the shyer of the two and is difficult to approach. Most often found singly or in pairs. Male sits on top of shrub calling. Favours low shrubland, particularly that dominated by acacia, saltbush and bluebush, often around wetlands and watercourses. Calls are best way to distinguish it from Chiming Wedgebill.

Varied Sittella ■ *Daphoenositta chrysoptera* 10–14cm

DESCRIPTION Variable species, with five races that vary in appearance. South-east mainland nominate race greyish above and white below with fine dark streaking, dark grey cap and dark tail with white tip. Upper wings dark and underwings have orange-rufous bar. Other races variously have white heads, striated breasts, black caps or white wings. **DISTRIBUTION** Found across mainland in all regions except central-east inland, and far north-west WA in Pilbara region to NT border. **HABITS AND HABITAT** Gregarious; usually seen in flocks, noisily and conspicuously moving briskly between trees or foraging hurriedly on trunk or over branches in all directions, even upside down. Found in a range of eucalypt woodland and forests, particularly rough-barked species. Decorates outside of nest to match tree bark.

Black-faced Cuckoo-shrike ■ *Coracina novaehollandiae* 29–36cm

DESCRIPTION Large cuckoo-shrike with blue-grey upperparts, and large black face-mask that continues down to throat. Grey breast blends to white underbelly. Sexes are alike. **DISTRIBUTION** Widespread across mainland and Tas. **HABITS AND HABITAT** Found singly or in twos. Forages mainly in tree canopy, or sometimes close to the ground, swooping down to capture prey. Distinctive habit of shuffling wings when landing on a perch, earning it the early common name of 'shufflewing'. Found in almost any wooded habitat across Australia, with the exception of rainforests. Also found in urban areas. Quite vocal at times, with contact calls in flight often first sign of bird.

Common Cicadabird ■ *Coracina tenuirostris* 24–25cm

DESCRIPTION Medium-small cuckoo-shrike. Adult male almost wholly dark blue-grey, with blackish face-mask and black wing feathers, edged grey. Female grey-brown above,

buff below with brown barring, and with fawn eyebrow and line under eye. **DISTRIBUTION** Coastal eastern and northern mainland, from Kimberley (WA) across Top End and down to north-east of Melbourne, although absent from lower Gulf of Carpentaria. **HABITS AND HABITAT** Found singly, or most often in twos. Quiet, unobtrusive and usually inhabits highest reaches of canopy. Presence usually noted via distinctive cicada-like call, and it gleans insect prey from leaves and branches. Mainly found in rainforests, paperbark woodland and mangroves, and in south in open eucalypt forest adjacent to primary habitat.

White-winged Triller ■ *Lalage tricolor* 17–19cm

DESCRIPTION Small member of cuckoo-shrike family. Breeding male pied, with glossy black head to below eye, continuing down nape and hindneck to back and tail, broken by

grey rump. Wings with white shoulder-patch and fringes to coverts; wholly white below. Eclipse male has rufous head and back. Female generally rufous where male is black. **DISTRIBUTION** Resident in northern and central mainland, and summer breeding migrant to south of mainland; absent from Tas. **HABITS AND HABITAT** Usually seen singly or in pairs, and at times occurs in larger flocks on migration to southern areas. Vocal during breeding, and often heard before it is seen, as males call flying from one tree to another in display. Insectivorous and forages in mid-upper canopy in lightly timbered eucalypt woodland and forest, often with grassy ground layer or sparse shrubs.

Crested Shrike-tit ■ *Falcunculus frontatus* 16–19cm

DESCRIPTION Unmistakable bird. In nominate race head has distinctive black-and-white striped pattern and black crest, and beak is short, deep and sharp. Upperbody mostly olive-green, and underparts vivid yellow. Throat black in male, olive-green in female. **DISTRIBUTION** Nominate race in south-east mainland; other two races in Top End of NT, and south-west WA. **HABITS AND HABITAT** Seen singly, in twos or in small parties. Found mostly in trees, where it occurs at all levels up to canopy. Sound of birds pulling off bark with the sharp beak is often heard before they are seen. Inhabits eucalypt forests and woodland, often along timbered watercourses, and also occurs in parks, gardens and farmland.

LEFT *Male;* RIGHT *Female*

Olive Whistler ■ *Pachycephala olivacea* 18–21cm

DESCRIPTION Medium-sized whistler with large, round, dark grey head. White throat mottled grey with grey breast-band, olive-brown upperparts and buff underparts. Female has lighter grey head than male.
DISTRIBUTION On and coastal of Great Dividing Range in south-east of mainland from southern Qld to western Vic., and in Tas. and Bass Strait islands. **HABITS AND HABITAT** Mostly seen singly unless breeding; shy and elusive species more often heard than seen. Forages in lower strata of habitat, and on fallen timber, logs and the ground in search of insects. Favours dense vegetation like understorey of rainforests and wet eucalypt forest, but also in coastal scrub and heathland.

Red-lored Whistler
■ *Pachycephala rufogularis* 19–22cm

DESCRIPTION Slender whistler with grey to olive-grey upperparts, rufous-orange lores, lower face and throat, grey breast-band and paler rufous-orange underparts. Female uniformly duller than male. **DISTRIBUTION** Range restricted to suitable habitat in area of north-west Vic. and adjoining south-east SA. Also in isolated pocket in central inland NSW. **HABITS AND HABITAT** Endangered species of mallee woodland associations, with requirement for long, unburnt habitat and mature trees. Also in cypress pine and Broombush. Found singly and in pairs, and very shy and hard to approach. Like other whistlers, usually heard before it is seen. Mainly forages on the ground for invertebrates, and makes a substantial cup-shaped nest on top of spinifex grass clump or low in mallee tree or shrub.

Golden Whistler ■ *Pachycephala pectoralis* 16–19cm

DESCRIPTION Distinctive; male bright buttercup-yellow underneath and around collar, with olive-green back and wings. Head black, and white throat separated from yellow chest by broad black band. Female grey above and paler grey below. **DISTRIBUTION** South-east mainland from Cairns region (Qld) to Eyre Peninsula (SA); south-west WA; Tas.; coastal islands. **HABITS AND HABITAT** Seen singly or in pairs. Forages in canopy of trees or within shrubs for insects. Found in almost any wooded habitat, from rainforest to mallee, but prefers denser habitats such as rainforests, eucalypt forest and woodland. Also visits gardens at times. Males call regularly.

LEFT Male; RIGHT Female

Rufous Whistler ■ *Pachycephala rufiventris* 15–18cm

DESCRIPTION Medium-sized whistler. Male has dark grey upperparts, and black face-mask and breast-band encircling white throat-patch. Rest of underparts rufous, which is paler on inland race. Female dull grey to brown with smutty-white throat-patch, and all underparts streaked darker. **DISTRIBUTION** Widely distributed across mainland, except driest desert regions; summer migrant to southern parts of range. **HABITS AND HABITAT** Seen singly or in pairs; occasionally in small groups on migration or post-breeding. During breeding season utters one of the most common calls in suitable habitat, with male giving loud, rollicking, slightly wheezy calls regularly. Gleans insects from foliage and branches of trees and shrubs in favoured habitat of eucalypt forest, woodland and shrubland particularly those with shrubby understorey

LEFT *Male*; RIGHT *Female*

Grey Shrike-thrush ■ *Colluricincla harmonica* 22–27cm

DESCRIPTION Medium-large, somewhat nondescript bird. Almost entirely grey with olive-grey back, and pale grey-white cheeks and underparts. Immatures take several years to attain full grey adult plumage, during which time they have variously rufous eyebrows and wing-coverts. **DISTRIBUTION** Found across mainland and Tas., Bass Strait islands. **HABITS AND HABITAT** Versatile and well-adapted species. Occurs in vast array of habitats, including eucalypt forest and woodland, coastal scrub, mallee and saltbush. Forages on the ground and within all strata in vegetation (logs, tree trunks, shrubs and canopy), mainly by gleaning invertebrates. Most often seen singly or in pairs; also in small family parties after breeding.

Crested Bellbird ■ *Oreoica gutturalis* 19–23cm

DESCRIPTION Distinctive bird; male has white face bordered by black crest above, which extends to a bib on breast. Vivid orange eyes and grey head. Body generally grey-brown, paler underneath.

Female lacks black on face and bib. **DISTRIBUTION** Found widely across arid and semi-arid inland regions, from central-west NSW and Vic. to WA coast, north to below tropical coastal band. **HABITS AND HABITAT** Heard well before it is seen, with male having an amazing loud call that is ventriloquial and frustratingly difficult to locate. Usually shy and unobstrusive, and females are seen much less often than males. Well known for habit of lining rim of nest with hairy caterpillars, thought to be for nest defence. Favours dry acacia shrubland and woodland, eucalypt woodland (especially mallee), and chenopod-spinifex associations.

Australasian Figbird ■ *Sphecotheres vieilloti* 26–29cm

DESCRIPTION Medium-sized member of oriole family. Male has black head surrounding large red patch of facial skin. Back olive-green, flight feathers and tail blackish, and underparts lime-yellow. Female mottled olive-green on upperparts including head, cream below with broad greenish streaks and small grey patch of facial skin. **DISTRIBUTION** Coastal northern and north-eastern mainland, down east coast to south-central NSW. Vagrant to Vic. **HABITS AND HABITAT** Gregarious and usually seen in flocks of 10–30 birds. Forages noisily on fruits, in particular figs, and not shy. Often found in fruiting trees in towns and gardens, but also in rainforests and adjacent eucalypt forest and woodland. Acrobatic forager; also known to swing upside down in rain shower to bathe before resuming foraging.

LEFT Male; RIGHT Female

Olive-backed Oriole
■ *Oriolus sagittatus* 25–30cm

DESCRIPTION Elegant, medium-sized oriole. Rich olive-green head, back and wings, darker flight feathers and tail, and white underparts with broad dark olive-green streaks. Red eyes and red-orange bill. Immatures paler than adults, and lack red eye and bill. **DISTRIBUTION** From Broome (WA), across northern and eastern Australia around to Adelaide (SA). **HABITS AND HABITAT** Most often seen singly or in twos, and known in north to associate with figbirds, catbirds and bowerbirds. Birds in southern mainland are summer migrants, arriving to breed in spring and departing in autumn. Almost strictly arboreal, searching foliage and branches for prey, which is often beaten before being consumed. Inhabits mainly eucalypt forest and woodland, but also rainforests. Distinctive and resonant call usually heard before bird is seen.

White-browed Woodswallow ■ *Artamus superciliosus* 18–20cm

DESCRIPTION Graceful-looking woodswallow. Male blue-grey above, with black face and breast, and distinctive white eyebrow. Underparts rich chestnut. Female lighter grey above and pale buff below, lacking black face. **DISTRIBUTION** Stronghold is eastern inland Australia, from NT and SA east to Great Dividing Range, but migrates to all areas except Cape York (Qld) and Tas. **HABITS AND HABITAT** Seasonal migrant capable of nomadic movements, heading south and west during summer months. Almost always seen with **Masked Woodswallow** *A. personatus* in mixed flocks; also forages with bee-eaters and honeyeaters, hawking insects from the air. Takes nectar occasionally. Mainly occurs in open eucalypt woodland, but also farmland and grassland sparsely populated with trees.

LEFT Male (top), Female (bottom); RIGHT Female

Dusky Woodswallow ■ *Artamus cyanopterus* 16–19cm

DESCRIPTION Medium-sized dark woodswallow. Head, neck and breast darkish brown, darker on back and belly. Black face and blue bill tipped with black. Wings dark gun-metal

grey-blue with thin white leading edge. Black tail tipped with white. Sexes are alike. **DISTRIBUTION** Eastern and southern mainland from Atherton Tableland (Qld) to south-west WA; also Tas. **HABITS AND HABITAT** Gregarious bird, usually in flocks of up to 30 birds. Noisy and conspicuous, capturing insect prey on the wing as it busily darts from tree to tree. Twists tail from side to side on landing. Also forages on the ground or among foliage occasionally. Inhabits open sclerophyll forest and woodland including mallee, but at times occurs in shrubland and heathland.

Grey Butcherbird ■ *Cracticus torquatus* 27–30cm

DESCRIPTION Small butcherbird. Black head, grey back and inner wings, and black outer wings and tail. Whitish throat, grey underparts and distinctive hooked-tipped bill. Sexes are alike. **DISTRIBUTION** Widespread across eastern and southern mainland, including

Tas.; also northern race in north-west WA-NT. **HABITS AND HABITAT** Carnivorous, taking insects, small birds, nestlings and eggs. Common name refers to habit of hanging prey on stick or branch, from where it can rip strips of flesh. Uses range of habitats, including eucalypt woodland, mallee, acacia shrubland, and commonly urban areas, where it can become quite tame if fed. Often mobbed by honeyeaters and other small birds to drive it out of area. Beautiful song uttered regularly at dawn and in the morning.

Australian Magpie ■ *Gymnorhina tibicen* 37–43cm

DESCRIPTION Black-and-white bird. Male has black head and underbody, white upperbody from nape down to upper tail, and white shoulder-patches. On female upperbody and shoulder-patches are grey rather than white. Several races, one of which is white on back. **DISTRIBUTION** Widely distributed across mainland, and north and east Tas. **HABITS AND HABITAT** Common and well-known species. Conspicuous and gregarious, and often in small groups. Intricate social structure and 'language', with much carolling and squabbling. Mostly forages on the ground, searching for insects and their larvae. Found wherever there are trees and adjacent open areas, including in urban parks. Tolerant of humans and readily become habituated and tame if fed. Notorious swooping behaviour during breeding season.

Pied Currawong
■ *Strepera graculina* 44–50cm

DESCRIPTION Distinctive large, mostly black bird, with small patches of white under tail, on tips and base of tail, and in shoulder of each wing. Large yellow eyes, massive black bill with small hook on end, and black legs. Sexes are alike. **DISTRIBUTION** Found on and in foothills of Great Dividing Range down to coastal areas of eastern mainland. **HABITS AND HABITAT** Usually seen singly, in twos or in small to large flocks. Forages at all levels from the ground to the tree canopy. Notorious nest predator for smaller birds, but also takes large amounts of insects and fruits. Inhabits eucalypt forest and woodland, and rainforests, and well adapted to suburban areas. Formerly seasonal migrant to cities like Sydney, but now largely resident due to planted fruiting trees like Cotoneaster and Privet.

Spangled Drongo
■ *Dicrurus bracteatus* 28–33cm

DESCRIPTION Medium-sized, wholly glossy black bird, with distinctive glossy green spots on head, neck and breast. Diagnostic flaring 'fishtail', large black bill with bristles at base and bright red eyes. Immature birds duller than adults and lack red eyes. **DISTRIBUTION** Tropical, subtropical and temperate northern and eastern Australia, to south coast NSW; vagrant to Vic. **HABITS AND HABITAT** Usually seen in singles or twos, but in larger flocks in non-breeding season. Often frantically active and noisy, and tolerant of people and close approach. Extremely acrobatic in pursuit of flying insects, which are also gleaned from foliage, and occasionally takes nectar. Typically found in rainforests, but also dry and wet sclerophyll forests and woodland, parks and gardens.

Rufous Fantail ■ *Rhipidura rufifrons* 14–18cm

DESCRIPTION Distinctive small bird. Large, fan-shaped tail that is fiery rufous over rump, then white-tipped on blackish feathers. Head, back and wings brown, rufous

eyebrow, curved white throat-stripe below blackish face, and black-and-white scalloped breast. **DISTRIBUTION** Coastal northern and eastern Australia, from Kimberley coast through NT, Qld and NSW to western Vic. **HABITS AND HABITAT** Active, darting bird of understorey of tropical and temperate rainforests and wet eucalypt forest, but also swamp woodland and mangroves in northern part of range. Usually seen singly or in pairs. Vocal and generally confiding of observers. Exhibits seasonal altitudinal movements where it can be found on passage in dry woodland and forests.

Grey Fantail ■ *Rhipidura fuliginosa* 14–16cm

DESCRIPTION Widespread and well-known fantail with several races. Generally grey above, with white eyebrow, and broad stripe on throat and edges of tail. Underparts largely buff-cream. Conspicuous whisker-like feathers around base of beak. Sexes are alike. **DISTRIBUTION** Widely across mainland, Tas. and coastal islands. **HABITS AND HABITAT** Mix of sedentary and migratory races; movements not well understood. However, Tasmanian birds known to cross Bass Strait and overwinter on mainland, returning in spring to breed. Typically seen singly or in twos. Conspicuous, active and noisy, and forages by fluttering about in the undergrowth and outer regions of mid-lower levels of canopy. Uses a wide range of habitats, although prefers eucalypt forest and woodland. Often in gardens and parks.

Willie Wagtail ■ *Rhipidura leucophrys* 18–21cm

DESCRIPTION Easily identified, small pied bird. Wholly black upperparts with fine white eyebrow, and white underparts start at neatly cut-off breast-line. Sexes are alike. **DISTRIBUTION** Common and widespread; found across whole of mainland and coastal islands. Rare vagrant to northern Tas. **HABITS AND HABITAT** Active and busy bird, typically seen chasing insects on the wing or darting around logs, shrubs and the ground in search of prey. Constantly fans and wags tail from side to side, and utters various calls. May call during still, moonlit nights. Favours open habitats, particularly open forests and woodland. Often associated with watercourses and wetlands; common in urban areas like parks, gardens and golf courses.

Australian Raven ▪ *Corvus coronoides* 46–53cm

DESCRIPTION Large, wholly glossy black bird with very large black bill, white eyes, and large patch of shaggy throat shackles that are longer than those of other corvids. Easiest

way to differentiate ravens and crows in Australia is by call; this species has a long, drawn-out and descending wail in final note of call. **DISTRIBUTION** Most widespread of Australian ravens, found across south-east and east Australia from south-east coast to Gulf of Carpentaria; also across southern coast to south-west WA. **HABITS AND HABITAT** Seen mostly singly or in pairs; easily confused with other ravens and crows. Occurs mainly in open but treed habitat like lightly timbered grassland; also modified environments like towns and rubbish tips, and roadsides on road kill. Noisy and conspicuous, although generally wary outside settlements.

Torresian Crow

▪ *Corvus orru* 46–51cm

DESCRIPTION Large, wholly black crow with large black bill and white eyes. Throat hackles present but not as pronounced as in Australian Raven (see above); no obvious mound or 'beard' from them. Like other Australian corvids, best identified by call; utters short, staccato *ugh-ugh-ugh* with a nasal quality, and last note often drawn out. **DISTRIBUTION** Across northern and inland Australia, from mid-south coast of WA to mid-north coast of NSW, except driest deserts and far south-west Qld. **HABITS AND HABITAT** On alighting lifts and shuffles wings in characteristic fashion. Mostly seen in singles, pairs or small flocks, and readily forages alongside **Little Crows** C. *bennetti*. Noisy and conspicuous, foraging mainly on the ground, and has adapted to humans by regularly using rubbish tips; also plains, paddocks and roadsides on road kill.

Leaden Flycatcher ■ *Myiagra rubecula* 14–17cm

DESCRIPTION Medium-sized, elegant woodland bird. Male has glossy blue-grey head, neck and upperparts, sharply demarcated from white underparts at breast. Female is greyer above with rufous-orange breast and throat. **DISTRIBUTION** Around northern and eastern Australia, from Kimberley's (WA) to western Vic. **HABITS AND HABITAT** Seasonal migrant in southern Australia, only present in spring and summer. Resident in northern Australia. Easily confused with **Satin Flycatcher** M. *cyanoleuca*, in which male is glossy dark blue-black above. Male gives sweet display calls, at the same time characteristically quivering tail. Insectivorous, and found in woodland including coastal scrub, eucalypt woodland and paperbarks. Nest a small cup on horizontal branch, decorated with lichen and bark.

LEFT *Male*; RIGHT *Female*

Restless Flycatcher
■ *Myiagra inquieta* 19–22cm

DESCRIPTION Medium-sized bird with glossy blue-black crown, face, back and wings, sharply demarcated from entirely white underparts. At times a slight orange-brown tint on breast, also present in immatures. Sexes are similar. **DISTRIBUTION** Across northern (except Cape York), eastern and southern Australia, to south-west WA. **HABITS AND HABITAT** Most often found singly or in pairs. Active, noisy and conspicuous, even when perched. Extremely mobile and able to hover while feeding; when doing so makes grinding call akin to someone sharpening scissors on grinding wheel. Bird of open eucalypt forest and woodland; northern birds also in paperbark woodland. Frequently seen on farmland.

White-eared Monarch

■ *Carterornis leucotis* 13–15cm

DESCRIPTION Distinctive pied monarch. Head black with white eyebrow, cheek-patch and scalloped throat. Upperparts grey-black with white edges to some coverts and primary feathers, and underparts wholly white from breast down. **DISTRIBUTION** Found in coastal region from tip of Cape York (Qld) to far north-east NSW. **HABITS AND HABITAT** Generally occurs in upper levels of rainforests, making close observation difficult, although it is active when foraging. Usually seen singly or in twos. Inhabits coastal and subcoastal rainforests, dry and wet sclerophyll forests, and ecotone between the two. Usually sallies for insect prey, and sits on tops of dead branches in wait.

Black-faced Monarch ■ *Monarcha melanopsis* 16–19cm

DESCRIPTION Medium-sized monarch. Grey head and breast; black face above bill to forehead, and below bill to throat; darker grey back, wings and tail. Underparts rufous-

orange from lower breast to undertail-coverts. **DISTRIBUTION** Coastal eastern mainland from tip of Cape York (Qld) to eastern Vic.; range spreading west towards Melbourne. **HABITS AND HABITAT** Usually seen in singles or twos. Summer migrant to eastern Australia, only being present in south of range in September–April each year. Favours rainforest habitat, but also occurs in wet sclerophyll forest, particularly where adjacent to rainforest. Forages for insects through mid-upper canopy layer, hopping from branch to branch like whistlers do, or sallying in mid-air.

Magpie-lark ■ *Grallina cyanoleuca* 24–30cm

DESCRIPTION Distinctive pied bird. Head, breast and upperparts largely black with prominent white eyebrow, neck-patch/line and line in closed wing. Underparts and rump white. Female has all-white face with no white eyebrow. **DISTRIBUTION** Across mainland and coastal islands; vagrant to Tas. and Bass Strait islands. **HABITS AND HABITAT** Common bird of urban areas usually seen singly, in twos or in small flocks, foraging on the ground on ovals, parks and gardens. Also wades at edges of dams, rivers and ponds. Natural habitats utilized include wide variety of open and lightly timbered areas and grassland, almost always near water. One of Australia's mud nesters, it makes a bowl-shaped nest from mud, feathers and grass.

White-winged Chough ■ *Corcorax melanorhamphos* 44–50cm

DESCRIPTION Distinctive large bird. Almost wholly black except for prominent red eye and large white wing-patch, visible particularly in flight. Strongly downcurved bill. Sexes are alike. **DISTRIBUTION** Across south-east Australia from central Qld coast to mid-Eyre coastline of SA. Absent from Tas. **HABITS AND HABITAT** Very active, noisy and gregarious species with a complex social structure. Occurs in sedentary, cooperatively breeding groups of up to 20 birds throughout the year. Usually seen on the ground, foraging for invertebrates in leaf litter or digging in the ground for bulbs and other vegetation. Utilizes forests and woodland near permanent water such as streams and farm dams. Mud-nesting species, with large, bowl-shaped nest made from mud on horizontal branch.

Victoria's Riflebird ■ *Ptiloris victoriae* 21–27cm

DESCRIPTION Stunning and unmistakable bird. Male mostly velvety-black with iridescent blue-green plumage on crown, breast and central tail. Underparts black with glossy coppery-green edging to feathers. Long, downcurved black bill. Female rufous-brown above and cream on throat, with rich buff underparts that contain brown chevrons and scalloping.
DISTRIBUTION North-east Qld, from near Cooktown south to Townsville region.

HABITS AND HABITAT All three species of riflebird in Australia are easily separated by range alone; this is the most northerly. Largely confined to tropical rainforests, but also in adjacent wet sclerophyll forest, mangroves and swampy woodland; usually seen singly or in twos. Arboreal, foraging on insects and fruits. Display of male spectacular, on top of branch or stump. Forms disc over head with wings, and flicks them from side to side while rocking head in between wings and displaying throat plumage.

Lemon-bellied Flycatcher
■ *Microeca flavigaster* 12–14cm

DESCRIPTION Small, pretty robin. Upperparts dull olive-green, wings and tail darker. Faint cream eyebrow, whitish throat and pale lemon underparts. Legs dark grey, in contrast to similar **Yellow-legged Flycatcher** M. *griseoceps*.
DISTRIBUTION Restricted to tropical zone across coastal northern Australia and islands, from Kimberley coast (WA) to central Qld coast. **HABITS AND HABITAT** Mostly seen singly or in twos. Typically noted perched on fence post, bare branch or overhead wire scanning the ground for insect prey to pounce on. Quiet and unobtrusive; generally tame and allows close approach. In WA and NT inhabits mangroves and rainforests, but in Qld favours open eucalypt forest and woodland. Also uses parks and gardens.

Red-capped Robin ■ *Petroica goodenovii* 11–13cm

DESCRIPTION Distinctive small robin. Male has conspicuous scarlet cap and breast, black upperparts with broad white wing-bar, and white belly and undertail. Female brown above with scarlet wash to cap; underparts creamy-buff. **DISTRIBUTION** Found across southern half of mainland, generally south of Tropic of Capricorn and west of Great Dividing Range. **HABITS AND HABITAT** Mostly seen singly or in pairs. Generally quiet and inconspicuous, although at times male is quite vocal. Forages mostly on or near the ground, pouncing on insect prey on the ground. Favours variety of dry habitats such as eucalypt, acacia and cypress pine woodlands, mallee, and modified areas like golf courses and orchards. Can be confused with **Scarlet** and **Flame Robins** P. *multicolor* and P. *phoenicea*, although these both lack scarlet cap.

LEFT Male; RIGHT Female

Rose Robin ■ *Petroica rosea* 11–13cm

DESCRIPTION Small robin. Male dark grey above with bright rose-red breast and white belly. Small white mark above bill. Female grey-brown above and creamy-buff below, often with some rose wash in breast. **DISTRIBUTION** Found around south-east mainland, on and coastal of Great Dividing Range, from Rockhampton (Qld) to Adelaide region (SA). **HABITS AND HABITAT** Usually seen singly or in pairs. Highly arboreal and actively forages through foliage in manner befitting a flycatcher. Seasonal altitudinal migrant, in high country only in spring–summer; coastally and on slopes of ranges in autumn–winter. Found in range of habitats, from rainforests and wet temperate forests in spring, to dry, open woodland in autumn and winter. Care needs to be taken to differentiate it from similar **Pink Robin** P. *rodinogaster*.

Hooded Robin ▪ *Melanodryas cucullata* 15–17cm

DESCRIPTION Large pied robin. Male has black hood, breast, back and wings with white shoulder-bar and wing-stripe. Underparts white; at times smutty on breast. Female similar to male, but grey-brown above, with grey breast and dirtier white underparts. **DISTRIBUTION** Found widely across mainland, except driest deserts and tropical Top End and Cape York, and coastal areas of Qld. Absent from Tas. **HABITS AND HABITAT** Typically seen in pairs or small groups. Generally shy and inconspicuous. Perches on low, exposed sites such as dead branches, tree stumps and fence posts, before pouncing onto the ground to capture insect prey. Inhabits range of lightly timbered woodland, particularly that dominated by acacias or eucalypts. Declining in regions of south-east Australia; sensitive to size of remnant habitat.

LEFT Male; RIGHT Female

Eastern Yellow Robin ▪ *Eopsaltria australis* 15–16cm

DESCRIPTION Medium-sized robin. Grey above with yellow patch on rump (olive-yellow in west of range), rich yellow below, and with off-white chin. Sexes are alike.

DISTRIBUTION Eastern and south-east mainland, from Cooktown (Qld) to south-east SA. **HABITS AND HABITAT** Sedentary species usually seen in pairs, or occasionally in small family parties post-breeding. Perches and pounces from low branches, or hangs from sides of tree trunks waiting for prey. Found in diverse array of habitats, including dry eucalypt woodland, wet sclerophyll forest, mulga, Brigalow, orchards and gardens. Generally quiet, although does tolerate close approach and can acclimatize to human disturbance in parks and gardens.

White-breasted Robin ■ *Quoyornis georgianus* 14–17cm

DESCRIPTION Medium-sized grey-and-white robin. Head and back darkish grey, blackish wings and tail, and off-white underparts. Faint whitish eyebrow, and white wing-bar visible in flight. Distinctive plumage. **DISTRIBUTION** Restricted to south-west WA, from Geraldton to Albany. **HABITS AND HABITAT** Most often seen singly or in pairs. Generally quiet and unobtrusive. Allows close approach as it sits in shrubs or on rocks, or hangs from trunks of trees in wait for insect prey to pounce on. Tends to favour dense habitats, such as coastal heaths and scrub in north of range, and understorey of dense forests in south.

Buff-sided Robin ■ *Poecilodryas cerviniventris* 14–17cm

DESCRIPTION Attractive medium-sized robin. Distinctive markings on head, with blackish face bordered above by broad white eyebrow and below by white throat and cheek-line. Upperparts olive-brown; underparts grey-white with rich buff flanks. **DISTRIBUTION** From gulf country of western Qld, across Top End to Kimberley region of WA. **HABITS AND HABITAT** Found in monsoon rainforest, pandanus-lined watercourses and swampy woodland. Usually occurs in pairs. Quietly works through habitat, pouncing on insects from exposed branches and rocks, and occasionally sallying for prey. Inquisitive and responds to pishing or imitations of calls. Only recently split from **White-browed Robin** *P. superciliosa* of north-east Qld, which lacks buff sides of this species.

Southern Scrub-robin ■ *Drymodes brunneopygia* 18–22cm

DESCRIPTION Distinctive long-tailed robin. Olive-brown above with white-edged black flight feathers and rufous-washed tail and rump. Underparts creamy-buff. Large dark eyes,

white partial eye-ring with black vertical line through eye, and whitish face. **DISTRIBUTION** Isolated population in central-west NSW mallee, then from north-west Vic. and south-west NSW across to south-west WA in suitable habitat. **HABITS AND HABITAT** Ground-dwelling species of mallee, Broombush and dry scrub; also some coastal heaths in south-west WA. Forages mostly on the ground, searching through soil and leaf litter for insects like beetles, and for spiders. Sits on top of exposed branch when singing. Seen singly or in pairs. Generally furtive and hard to approach, but at times inquisitive. Rarely flies.

Horsfield's Bushlark ■ *Mirafra javanica* 12–15cm

DESCRIPTION Small, somewhat thickset lark. Variable, with numerous subspecies, but generally has scalloped rufous-and-brown upperparts, buff eyebrow, and cream-buff underparts with broad streaking on breast. Smaller than similar Australasian Pipit (see p. 151) and **Eurasian Skylark** *Alauda arvensis* (introduced). **DISTRIBUTION** Found across

northern, eastern and south-eastern mainland, inland as far as Lake Eyre basin of SA. **HABITS AND HABITAT** Usually seen singly or in pairs; can be in large flocks outside breeding season. Cryptic and well camouflaged, and easily overlooked. Inhabits range of native and introduced grasslands, including pastures on farmland. Impressive mimic, incorporating songs of other species into its own. Male displays in song flight, hovering above the ground on quivering wings; also sits on top of fence posts or stumps to sing.

Golden-headed Cisticola ■ *Cisticola exilis* 9–11cm

DESCRIPTION Small warbler. In breeding condition male has golden-orange head with small, erectile crest, and golden wash to neck and flanks. Back and wings blackish with rufous edges to feathers, and creamy-buff underparts. Non-breeding male and female have black streaking on crown.
DISTRIBUTION Coastal northern, eastern and south-eastern mainland; vagrant to Tas. **HABITS AND HABITAT** Busy and active bird, constantly on the move through rank grasses, reeds and similar vegetation around water. Male calls regularly from exposed reeds, with crest erect. Also performs display flights, circling upwards jerkily while singing, before descending rapidly into cover. Usually seen singly or in pairs. Male approachable in breeding season, but otherwise quiet and unobtrusive. Easily separated from similar **Zitting Cisticola** *C. juncidis* by call.

Australian Reed-Warbler ■ *Acrocephalus australis* 15–18cm

DESCRIPTION Sleek and plain appearance. Darkish olive-brown upperparts and fawn-buff underparts, rump and faint eyebrow. **DISTRIBUTION** Widespread across eastern mainland and Tas., and coastal WA.
HABITS AND HABITAT Summer breeding migrant in south-east and south-west Australia. Beautiful singer, with one of the most common sounds of reed beds in warm months. Some birds overwinter in south, rarely calling. Generally cryptic and difficult to approach; presence almost always given away by male singing. Found singly or in pairs, and in small flocks at times. Hops through and up reeds in search of insects, and sings from top of sturdy reed. Favours reeds, cumbungi and similar along watercourses and around wetlands. Nest an impressive cup woven around reed stems.

Rufous Songlark
■ *Megalurus mathewsi* 19–21cm

DESCRIPTION Grey-brown above, with pale edging to brown flight feathers; underparts off-white with very pale brown breast. Faint white eyebrow, dark line through eye, and distinctive rufous rump and uppertail-coverts. Breeding male has black bill and mouth; pale in female and non-breeding male. Male also larger than female. **DISTRIBUTION** Widespread across mainland, except Cape York, dense wet forests and driest deserts. **HABITS AND HABITAT** Migratory species to southern Australia in spring and summer. Suitable habitat resounds with its calls after arrival from northern regions. Male flies from perch to perch on slightly quivering wings, calling loudly. Found in range of habitats, but favours open, grassy woodland and scrub. Nest a neat, grass-lined cup on the ground, usually tucked under fallen timber.

Silvereye ■ *Zosterops lateralis* 11–13cm

DESCRIPTION Variable, but commonly with olive-green head, conspicuous ring of white feathers around eye, and pale yellow-green chin and throat. Back grey and wings olive-

green, with pale-buff flanks and white undertail. **DISTRIBUTION** Eastern and southern Australia, coastal islands including Great Barrier Reef, Bass Strait and Tas. **HABITS AND HABITAT** Almost always seen in small flocks. Calls regularly and usually quite conspicuous despite small size. Forages actively in vegetation, gleaning from leaves and branches; very occasionally forages on the ground. Often found in areas with small honeyeaters and thornbills. Inhabits mostly wooded areas, especially where there is a good shrub layer, and also uses orchards, urban parks and gardens.

Welcome Swallow ■ *Hirundo neoxena* 14–16cm

DESCRIPTION The familiar 'house' swallow of Australia. Metallic blue-black above, with rusty-reddish forehead, throat and upper breast. Lower breast and belly light grey. Tail long and deeply forked, with a series of small white spots. Sexes are alike. **DISTRIBUTION** Eastern and inland Australia to central NT, southern Australia to south-west WA; Tas. Scarce in some other regions. **HABITS AND HABITAT** Gregarious; most often seen in flocks of up to 30 birds. Active and catches prey in flight using acrobatic flying skills. Also seen perched on overhead wires, fences and bare branches. Builds half-cup mud nest under rock ledges, building roofs, garages and sheds. Inhabits wide variety of habitats, except dense forests and dry inland areas, and commonly found in towns and cities.

Fairy Martin

■ *Petrochelidon ariel* 12–13cm

DESCRIPTION Small, short-tailed swallow. Rufous crown and nape, glossy blue-black back, wings and tail, and white rump. Underparts whitish with fine black streaks. Similar **Tree Martin** *P. nigricans* has blue-black head and dirtier white rump. **DISTRIBUTION** Widespread across mainland, except driest deserts. Vagrant to Tas. **HABITS AND HABITAT** Also known as Bottle Swallow. Nests colonially in bottle-shaped mud nests hung under cliffs, on walls, and under artificial structures like bridges and road culverts. Often found in mixed flocks with Tree Martin and other swallows. Spends most of its time on the wing in search of flying insects. Found over open habitats like wetlands and rivers in vicinity of nest sites.

Bassian Thrush ■ *Zoothera lunulata* 26–29cm

DESCRIPTION Large thrush. Olive-brown above with much black scalloping, buff-brown throat and whitish underparts scalloped black. Pale buff lores and eye-ring, and pink legs.

Separated from near-identical but smaller **Russet-tailed Thrush** *Z. heinei* mainly by calls. **DISTRIBUTION** Coastal eastern Australia in disjunct distribution from Atherton Tableland (Qld) to Mt Lofty Ranges (SA); also Tas. **HABITS AND HABITAT** Usually seen singly or in twos (probably pairs). Inhabits wet sclerophyll forest and rainforests, particularly densely vegetated gullies. Forages on insects in leaf litter; largely terrestrial and cryptic, with presence usually detected by calls. Around picnic grounds can become fairly tame, but otherwise quite timid.

Mistletoebird ■ *Dicaeum hirundinaceum* 10cm

DESCRIPTION Male has glossy blue-black head, wings and upperparts, bright red throat and breast, white belly with central dark streak, and bright red undertail. Female grey above, white below, with grey streak on belly, and paler, red-striped undertail. **DISTRIBUTION** Across mainland Australia and many coastal islands, although absent from Tas. Does not occur in driest treeless deserts of interior. **HABITS AND HABITAT** Occurs singly or in pairs. Usually seen high in tree canopy, or flying swiftly and erratically. Calls often. Uses wide variety of wooded habitats where mistletoe grows, especially eucalypt forest and woodland. Nest a delicate and soft hanging structure that resembles a baby's sock.

LEFT Male; RIGHT Female

Olive-backed Sunbird ■ *Cinnyris jugularis* 11–12cm

DESCRIPTION Small, honeyeater-like bird. Male olive-green above with thin yellow eyebrow and bright, glossy blue-black throat and breast. Remainder of underparts bright yellow; long, downcurved bill. Female lacks blue-black throat and breast. DISTRIBUTION North-east Qld around coastal areas of Cape York south to Bundaberg region. HABITS AND HABITAT The only sunbird in Australia; distinctive and easily identified. Inhabits woodland, gardens and rainforest edges where there are nectar-producing flowers in abundance. Usually in band within 25km of coast. Nest a large, hanging, pendulous structure with hooded side entrance, suspended from vines and saplings, or from houses and sheds in towns.

LEFT Male; RIGHT Female

Zebra Finch ■ *Taeniopygia guttata* 9–11cm

DESCRIPTION Distinctive. Male grey-brown above; face has large orange cheek-patch, and black-and-white markings. Flanks orange spotted with white, throat grey-barred black, and tail black with white barring. Female lacks orange in cheeks and flanks, and black throat barring. DISTRIBUTION Mainly across arid and semi-arid inland Australia, with some coastal movements during severe drought. HABITS AND HABITAT Found near water in arid rangeland habitats like mulga, gibber, woodland and open scrub; also in orchards and gardens. Gregarious; usually seen in small flocks, often perched on fences. Very active, foraging busily for seeds on the ground. Nest an untidy dome of grass, usually in shrub but often in hollow, fence post or abandoned Fairy Martin (see p. 145) nest.

Double-barred Finch ■ *Taeniopygia bichenovii* 10–12cm

DESCRIPTION Unmistakable small finch. Large white facial disk with thin black border, and black band across breast on white underparts. Upperparts grey-brown, wings black with delicate white spots, and tail black.

Rump white or black depending on race. **DISTRIBUTION** Northern and eastern mainland and coastal islands, from Kimberley region (WA) across Top End, and as far south as NSW- Vic. border. **HABITS AND HABITAT** Gregarious, even when breeding, typically in flocks of 10–20 birds. Often found in association with other finches, particularly when foraging, and can become quite tame in proximity to human activity. Favours open, grassy woodland, acacia woodland and sedgeland, and often occurs near wetlands like swamps and rivers; also on farmland and in towns.

Masked Finch ■ *Poephila personata* 11–14cm

DESCRIPTION Large yellow bill and black face-mask; salmon-brown above and paler below, with black band on flanks, white undertail and rump, and black tail. North Qld race

has white cheeks. **DISTRIBUTION** Nominate race across northern Australia, from Broome (WA) to Gulf of Carpentaria (Qld); another race on Cape York (Qld). **HABITS AND HABITAT** Gregarious; usually in small flocks that can concentrate in hundreds of birds in late dry season. Often mixes with species such as similar **Long-tailed Finch** *P. acuticauda*, Gouldian Finch (see p. 151) and other granivores like doves. Favours savannah woodland and grassland, particularly with dense grass layer. Usually found near water, and regularly visits each morning to drink. Generally approachable.

Plum-headed Finch

■ *Neochmia modesta* 10–12cm

DESCRIPTION Unmistakable. Rich plum-coloured crown and chin, and whitish underparts with broad buff-brown barring. Upperparts brown with white spots in wings and striping at base of tail. **DISTRIBUTION** Inland eastern Australia, through Qld and NSW, generally west of Great Dividing Range but coastal in central Qld. **HABITS AND HABITAT** Generally associated with wetlands and streams, where it favours tall grasses, reeds and cumbungi-fringed waterbodies. Gregarious, roosting together in reeds and similar over water; descends rapidly into reeds to roost communally. Furtive and difficult to approach. Forages on the ground on seeds. Often mixed with other finches, such as Zebra Finch and Double-barred Finch (see p. 147 and opposite).

Diamond Firetail ■ *Stagonopleura guttata* 11–13cm

DESCRIPTION Stocky, large-headed finch. Distinctive broad black breast-band and white-spotted flanks within white underparts. Red bill and eyes, grey head, ash-brown upperparts and vibrant crimson rump.

DISTRIBUTION South-east mainland, broadly from south-central Qld to south-east SA. **HABITS AND HABITAT** Occurs singly or in pairs during breeding season, and in small flocks in non-breeding season. Forages unobtrusively on the ground, usually in close proximity to trees or other vegetation. Plaintive, drawn-out but rising *grheee* call often heard before bird is seen. Inhabits open, grassy woodland, heath and farmland, or grassland with scattered trees. Nest a bulky grass dome in fork of tree, shrubs or mistletoe, but often in base of nest of raptor such as Whistling Kite or Wedge-tailed Eagle (see pp. 41 and 43).

Red-eared Firetail ■ *Stagonopleura oculata* 11–13cm

DESCRIPTION Upperparts and breast grey-brown with fine blackish barring; underparts black with bold white spots giving broadly barred appearance. Red bill, black face-mask,

blue eye-ring, and crimson red ear-coverts. Rump and lower tail crimson red. DISTRIBUTION Restricted to coastal south-west WA. HABITS AND HABITAT Found singly or in pairs. Mostly quiet and shy. Inhabits undergrowth in forests, dense riparian vegetation, coastal scrub and heathland. Easily overlooked, but can be inquisitive when observer imitates its call, or pishes. Forages on seeds of grasses and sedges, but also takes insects, particularly when breeding and feeding nestlings. Nest a long, flattened dome in foliage; lays 4–5 eggs.

Painted Finch ■ *Emblema pictum* 10–12cm

DESCRIPTION Stunning bird. Male has brown upperparts with red rump and black tail. Face red, with red continuing in fading line down breast to black underparts, which are spotted white on flanks and red in centre. White eyes. Female has brown face and only white spotting on underside. DISTRIBUTION Across arid inland and western mainland, from western Qld and NSW to WA coast.

HABITS AND HABITAT Closely associated with spinifex grass on rocky hills and escarpments near permanent water such as springs and waterholes, but also in acacia scrub and sand plains in some regions. Forages on spinifex seeds and other grasses, and usually seen in pairs or small flocks. Typically nests on top of spinifex tussock.

LEFT Male; RIGHT Female

Gouldian Finch
■ *Erythrura gouldiae* 12–14cm

DESCRIPTION Polymorphic; most birds have black head, about a quarter have red head and a few are golden headed. Head colour fringed in blue, rich grass-green upperparts, vivid purple breast, yellow belly and white vent. Blue rump, and black tail with two long, fine-ended central feathers. **DISTRIBUTION** Patchily across northern mainland, from near Broome (WA) to western side of Cape York (Qld). **HABITS AND HABITAT** Common aviary bird that has declined substantially in the wild due to inappropriate grazing and fire regimes. Uses savannah woodland, and favours areas with spear-grass *Sorghum* sp. in close proximity to water. Nest made inside hollows in eucalypts, which are also threatened by fire used to manage grasses for domestic stock. Subject to long-running recovery effort.

Australasian Pipit
■ *Anthus australis* 15–18cm

DESCRIPTION Mostly brown above with bold darker streaking, and buff-whitish below streaked on breast and flanks. Creamy-white eyebrows, curving lower cheek-stripe and throat. Tail dark brown with white edges. Sexes are alike. **DISTRIBUTION** Common and widespread across mainland and Tas. **HABITS AND HABITAT** Seen singly, but sometimes in twos or small flocks. Usually cryptic, perching unobtrusively on the ground, rocks, stumps and fence posts. During courtship male makes spectacular display flights, rising and falling in broad loops well above the ground while calling constantly. Occupies large range of open habitats with few or no trees, including clearings in open woodland, and is a common bird of farmland.

FURTHER INFORMATION

There exists a wealth of information on Australia's birds in a range of formats, from books and regional guides, to audio recordings and websites. The following is a list of key resources available to birders visiting Australia.

BIRDWATCHING ORGANIZATIONS

BirdLife Australia

Website: www.birdlife.org.au

Email: info@birdlife.org.au

The oldest conservation organization in Australia, first established as the Royal Australasian Ornithologists Union in 1901, it has 11,000 members in branches across the country. It also has a large research and conservation team striving to implement recovery actions for a range of threatened species, and runs the Atlas of Australian Birds under the banner of Birdata (see below).

SIGHTINGS DATABASES

Birdata

http://birdata.com.au

The online interface of BirdLife Australia's long-running Atlas programme, which has had over 18 million records entered since 1998. The data is stringently vetted and is now being used to produce Australia's first national bird indices, to inform policy makers and land managers.

eBird

http://ebird.org/home

Managed by the Cornell Lab of Ornithology, eBird is the world's largest biodiversity-focussed citizen science project and has over 100 million bird sightings submitted annually around the world. eBird data are freely accessible to anyone, and the website provides a great way to find out what has been seen recently at local 'hotspots'.

INTERNET

Facebook

In recent years there has been a surge in the use of Facebook for communicating on birding in Australia, including pages like 'Australian Bird Identification', 'Australian Twitchers', 'Victorian Birders', 'NSW & ACT birders' and 'South-east Queensland birders'. They provide a wealth of information and can connect visiting birders with local information or experts.

Eremaea

www.eremaea.com

Repository for sightings of significance, sorted and displayed by state or territory.

Feathers and Photos

www.feathersandphotos.com.au

Photography forum containing a vast array of Australian species, and also packed with information on identification, trip reports and equipment.

APPS

Pizzey & Knight Birds of Australia Digital Edition
The Michael Morcombe and David Stewart eGuide to Australian Birds
Both of these apps are available for Apple and Android devices, and they are the mobile version of their popular field guides. The advantage is that calls for nearly all species, and many of the subspecies, can be carried with you into the field. The Pizzey & Knight guide is also available in a PC version.

RECOMMENDED READING

Field guides
Menkhorst, P., Rogers, D., Clarke, R., Davies, J., Marsack, P., & Franklin, K. (2019). *The Australian Bird Guide*, 2nd edn. CSIRO Publishing, Clayton South.

Pizzey, G., Knight, F. & Pizzey, S. (2012). *The Field Guide to the Birds of Australia*, 9th edn. Harper Collins Publishers, Sydney.

Simpson, K. & Day, N. (2010). *Field Guide to the Birds of Australia*, 8th edn. Penguin Group (Australia), Camberwell.

Slater, P., Slater, P. & Slater, R. (2009). *The Slater Field Guide to Australian Birds*, 2nd edn. New Holland Publishers (Australia), Sydney.

Regional guides
Dolby, T. & Clarke, R. (2014). *Finding Australian Birds – a Field Guide to Birding Locations*. CSIRO Publishing, Collingwood.

Dolby, T., Johns, P. & Symonds, S. (2009). *Where to see Birds in Victoria*. Jacana Books, Crows Nest.

Neilsen, L. (2004). *Birding Australia*. Lloyd Neilsen, Mount Molloy.

Thomas, R., Thomas, S., Andrew, D. & McBride, A. (2011). *The Complete Guide to Finding the Birds of Australia*, 2nd edn. CSIRO Publishing, Collingwood.

Tzaros, C. (2005). *Wildlife of the Box-Ironbark country*. CSIRO Publishing, Collingwood.

References
Marchant, S. & Higgins, P. (eds). (1990). *Handbook of Australian, New Zealand and Antarctic Birds. Volume 1: Ratites to Ducks*. Oxford University Press, Melbourne.

Marchant, S. & Higgins, P. (eds). (1993). *Handbook of Australian, New Zealand and Antarctic Birds. Volume 2: Raptors to Lapwings*. Oxford University Press, Melbourne.

Higgins, P. & Davies, S. (eds). (1996). *Handbook of Australian, New Zealand and Antarctic Birds. Volume 3: Snipe to Pigeons*. Oxford University Press, Melbourne.

Higgins, P. (ed). (1999). *Handbook of Australian, New Zealand and Antarctic Birds. Volume 4: Parrots to Dollarbird*. Oxford University Press, Melbourne.

Higgins, P., Peter, J. & Steele, W. (2001). *Handbook of Australian, New Zealand and Antarctic Birds. Volume 5: Tyrant-flycatchers to Chats*. Oxford University Press, Melbourne.

Higgins, P. & Peter, J. (eds). (2002). *Handbook of Australian, New Zealand and Antarctic Birds. Volume 6: Pardalotes to Shrike-thrushes*. Oxford University Press, Melbourne.

Higgins, P., Peter, J. & Cowling, S. (2006). *Handbook of Australian, New Zealand and Antarctic Birds. Volume 7: Boatbill to Starlings*. Oxford University Press, Melbourne.

AUSTRALIAN BIRD LIST

Species level taxonomy follows IOC version 9.2 (Gill & Donsker, 2019), while common names follow nomenclature of *The BirdLife Australia Working List of Australian Birds* version 3.0 (BirdLife Australia, 2019).

Definitions of Australian status:

E   Endemic species; found only in Australia.
BE Breeding endemic; breeds only in Australia, but distribution not restricted to Australia.
R   Resident breeding Australian species, but not necessarily resident all year and may breed outside Australia (for example found in or migrates to/from Papua New Guinea or Indonesia).
X   Extinct; formerly wild resident.
XA Extinct in Australia, but still present elsewhere.
M   Migrant, passage migrant and non-breeding visitor, for example migratory waders that breed in Siberia.
V   Vagrant; species visiting Australia that is outside its normal range, but apparently in a wild state.
I   Introduced; occurs in wild state but not native to Australia and introduced from another region.

Abbreviations of IUCN Red List status:

CR Critically Endangered
EN Endangered
VU Vulnerable
NT Near Threatened
LC Least Concern

Common Name	Scientific name	Status	IUCN
Struthionidae (Ostriches)			
Ostrich	*Struthio camelus*	I	LC
Casuariidae (Cassowaries)			
Southern Cassowary	*Casuarius casuarius*	R	VU
Dromaiidae (Emus)			
Emu	*Dromaius novaehollandiae*	E	LC
Megapodiidae (Megapodes)			
Australian Brush-turkey	*Alectura lathami*	E	LC
Malleefowl	*Leipoa ocellata*	E	VU
Orange-footed Scrubfowl	*Megapodius reinwardt*	R	LC
Numididae (Guineafowl)			
Helmeted Guineafowl	*Numida meleagris*	I	LC
Odontophoridae (New World Quail)			
California Quail	*Callipepla californica*	I	LC
Phasianidae (Pheasants, Fowl and Allies)			
Wild Turkey	*Meleagris gallopavo*	I	LC
Stubble Quail	*Coturnix pectoralis*	E	LC

Common Name	Scientific name	Status	IUCN
Brown Quail	Coturnix ypsilophora	R	LC
King Quail	Excalfactoria chinensis	R	LC
Red Junglefowl	Gallus gallus	I	LC
Green Junglefowl	Gallus varius	I	LC
Common Pheasant	Phasianus colchicus	I	LC
Indian Peafowl	Pavo cristatus	I	LC
Anseranatidae (Magpie Goose)			
Magpie Goose	Anseranas semipalmata	R	LC
Anatidae (Ducks, Geese and Swans)			
Spotted Whistling-Duck	Dendrocygna guttata	R	LC
Plumed Whistling-Duck	Dendrocygna eytoni	R	LC
Wandering Whistling-Duck	Dendrocygna arcuata	R	LC
Cape Barren Goose	Cereopsis novaehollandiae	E	LC
Canada Goose	Branta canadensis	I	LC
Black Swan	Cygnus atratus	E	LC
Mute Swan	Cygnus olor	I	LC
Freckled Duck	Stictonetta naevosa	D	LC
Raja Shelduck	Tadorna radjah	R	LC
Australian Shelduck	Tadorna tadornoides	E	LC
Paradise Shelduck	Tadorna variegata	V	LC
Pink-eared Duck	Malacorhynchus membranaceus	E	LC
Australian Wood Duck	Chenonetta jubata	E	LC
Cotton Pygmy-goose	Nettapus coromandelianus	R	LC
Green Pygmy-goose	Nettapus pulchellus	R	LC
Garganey	Spatula querquedula	M	LC
Australasian Shoveler	Spatula rhynchotis	R	LC
Northern Shoveler	Spatula clypeata	V	LC
Eurasian Wigeon	Mareca penelope	V	LC
Mallard	Anas platyrhynchos	I	LC
Pacific Black Duck	Anas superciliosa	R	LC
Grey Teal	Anas gracilis	R	LC
Chestnut Teal	Anas castanea	E	LC
Northern Pintail	Anas acuta	V	LC
Eurasian Teal	Anas crecca	V	LC
Hardhead	Aythya australis	R	LC
Blue-billed Duck	Oxyura australis	E	NT
Musk Duck	Biziura lobata	E	LC
Podargidae (Frogmouths)			
Marbled Frogmouth	Podargus ocellatus	R	LC
Papuan Frogmouth	Podargus papuensis	R	LC
Tawny Frogmouth	Podargus strigoides	E	LC
Caprimulgidae (Nightjars)			
Spotted Nightjar	Eurostopodus argus	R	LC
White-throated Nightjar	Eurostopodus mystacalis	R	LC
Grey Nightjar	Caprimulgus jotaka	V	LC
Large-tailed Nightjar	Caprimulgus macrurus	R	LC
Savanna Nightjar	Caprimulgus affinis	V	LC
Aegothelidae (Owlet-nightjars)			
Australian Owlet-nightjar	Aegotheles cristatus	R	LC
Apodidae (Swifts and Swiftlets)			
Glossy Swiftlet	Collocalia esculenta	R	NT
Christmas Island Swiftlet	Collocalia natalis	E	LC
Australian Swiftlet	Aerodramus terraereginae	R	LC
Uniform Swiftlet	Aerodramus vanikorensis	V	LC
Papuan Spine-tailed Swift	Mearnsia novaeguineae	V	LC
White-throated Needletail	Hirundapus caudacutus	M	LC
Silver-backed Needletail	Hirundapus cochinchinensis	V	LC
Common Swift	Apus apus	V	LC

Common Name	Scientific name	Status	IUCN
Fork-tailed Swift	Apus pacificus	M	LC
House Swift	Apus nipalensis	V	LC
Otididae (Bustards)			
Australian Bustard	Ardeotis australis	R	LC
Cuculidae (Cuckoos)			
Lesser Coucal	Centropus bengalensis	V	LC
Pheasant Coucal	Centropus phasianinus	R	LC
Asian Koel	Eudynamys scolopaceus	R	LC
Eastern Koel	Eudynamys orientalis	R	LC
Pacific Long-tailed Cuckoo	Urodynamis taitensis	V	LC
Channel-billed Cuckoo	Scythrops novaehollandiae	R	LC
Horsfield's Bronze-Cuckoo	Chrysococcyx basalis	BE	LC
Black-eared Cuckoo	Chrysococcyx osculans	BE	LC
Shining Bronze-Cuckoo	Chrysococcyx lucidus	R	LC
Little Bronze-Cuckoo	Chrysococcyx minutillus	R	LC
Pallid Cuckoo	Cacomantis pallidus	BE	LC
Chestnut-breasted Cuckoo	Cacomantis castaneiventris	R	LC
Fan-tailed Cuckoo	Cacomantis flabelliformis	R	LC
Brush Cuckoo	Cacomantis variolosus	R	LC
Large Hawk-Cuckoo	Hierococcyx sparverioides	V	LC
Hodgson's Hawk-Cuckoo	Hierococcyx nisicolor	V	LC
Indian Cuckoo	Cuculus micropterus	V	LC
Oriental Cuckoo	Cuculus optatus	M	LC
Columbidae (Pigeons and Doves)			
Rock Dove	Columba livia	I	LC
White-headed Pigeon	Columba leucomela	E	LC
Oriental Turtle Dove	Streptopelia orientalis	V	LC
Barbary Dove	Streptopelia roseogrisea	I	LC
Red Turtle Dove	Streptopelia tranquebarica	V	LC
Spotted Dove	Spilopelia chinensis	I	LC
Laughing Dove	Spilopelia senegalensis	I	LC
Brown Cuckoo-Dove	Macropygia phasianella	E	LC
Common Emerald Dove	Chalcophaps indica	R	NT
Pacific Emerald Dove	Chalcophaps longirostris	R	LC
Common Bronzewing	Phaps chalcoptera	E	LC
Brush Bronzewing	Phaps elegans	E	LC
Flock Bronzewing	Phaps histrionica	E	LC
Crested Pigeon	Ocyphaps lophotes	E	LC
Spinifex Pigeon	Geophaps plumifera	E	LC
Squatter Pigeon	Geophaps scripta	E	LC
Partridge Pigeon	Geophaps smithii	E	VU
Wonga Pigeon	Leucosarcia melanoleuca	E	LC
Chestnut-quilled Rock-Pigeon	Petrophassa rufipennis	E	LC
White-quilled Rock-Pigeon	Petrophassa albipennis	E	LC
Diamond Dove	Geopelia cuneata	E	LC
Peaceful Dove	Geopelia placida	R	LC
Bar-shouldered Dove	Geopelia humeralis	R	LC
Nicobar Pigeon	Caloenas nicobarica	V	LC
Black-banded Fruit-Dove	Ptilinopus alligator	E	LC
Wompoo Fruit-Dove	Ptilinopus magnificus	R	LC
Superb Fruit-Dove	Ptilinopus superbus	R	LC
Rose-crowned Fruit-Dove	Ptilinopus regina	R	LC
Orange-bellied Fruit-Dove	Ptilinopus iozonus	V	LC
Orange-fronted Fruit-Dove	Ptilinopus aurantiifrons	V	LC
Elegant Imperial-Pigeon	Ducula concinna	V	LC
Christmas Imperial-Pigeon	Ducula whartoni	E	NT
Collared Imperial-Pigeon	Ducula mullerii	R	LC
Pied Imperial-Pigeon	Ducula bicolor	V	LC

Common Name	Scientific name	Status	IUCN
Torresian Imperial-Pigeon	*Ducula spilorrhoa*	R	LC
Zoe's Imperial-Pigeon	*Ducula zoeae*	V	LC
Topknot Pigeon	*Lopholaimus antarcticus*	E	LC
Rallidae (Crakes, Rails and Swamphens)			
Red-necked Crake	*Rallina tricolor*	R	LC
Red-legged Crake	*Rallina fasciata*	V	LC
Lord Howe Woodhen	*Gallirallus sylvestris*	E	EN
Buff-banded Rail	*Gallirallus philippensis*	R	LC
Lewin's Rail	*Lewinia pectoralis*	R	LC
Slaty-breasted Rail	*Lewinia striata*	V	LC
Corn Crake	*Crex crex*	V	LC
Pale-vented Bush-hen	*Amaurornis moluccana*	R	LC
White-breasted Waterhen	*Amaurornis phoenicurus*	R	LC
Baillon's Crake	*Porzana pusilla*	R	LC
Australian Spotted Crake	*Porzana fluminea*	E	LC
Ruddy-breasted Crake	*Porzana fusca*	V	LC
Spotless Crake	*Porzana tabuensis*	R	LC
White-browed Crake	*Porzana cinerea*	R	LC
Chestnut Rail	*Eulabeornis castaneoventris*	R	LC
Watercock	*Gallicrex cinerea*	V	LC
Australasian Swamphen	*Porphyrio melanotus*	R	LC
Common Moorhen	*Gallinula chloropus*	V	LC
Dusky Moorhen	*Gallinula tenebrosa*	R	LC
Lesser Moorhen	*Paragallinula angulata*	V	LC
Black-tailed Native-hen	*Tribonyx ventralis*	E	LC
Tasmanian Native-hen	*Tribonyx mortierii*	E	LC
Eurasian Coot	*Fulica atra*	R	LC
Gruidae (Cranes)			
Sarus Crane	*Grus antigone*	R	LC
Brolga	*Grus rubicunda*	R	LC
Podicipedidae (Grebes)			
Tricolored Grebe	*Tachybaptus tricolor*	V	LC
Australasian Grebe	*Tachybaptus novaehollandiae*	R	LC
Hoary-headed Grebe	*Poliocephalus poliocephalus*	E	LC
Great Crested Grebe	*Podiceps cristatus*	R	LC
Phoenicopteridae (Flamingoes)			
Greater Flamingo	*Phoenicopterus roseus*	V	LC
Turnicidae (Button-quail)			
Red-backed Button-quail	*Turnix maculosus*	R	LC
Black-breasted Button-quail	*Turnix melanogaster*	E	NT
Chestnut-backed Button-quail	*Turnix castanotus*	E	LC
Buff-breasted Button-quail	*Turnix olivii*	E	EN
Painted Button-quail	*Turnix varius*	R	LC
Red-chested Button-quail	*Turnix pyrrhothorax*	E	LC
Little Button-quail	*Turnix velox*	E	LC
Burhinidae (Stone-curlews)			
Bush Stone-curlew	*Burhinus grallarius*	R	LC
Beach Stone-curlew	*Esacus magnirostris*	R	LC
Chionididae (Sheathbills)			
Black-faced Sheathbill	*Chionis minor*	R	VU
Haematopodidae (Oystercatchers)			
South Island Pied Oystercatcher	*Haematopus finschi*	V	LC
Australian Pied Oystercatcher	*Haematopus longirostris*	E	LC
Sooty Oystercatcher	*Haematopus fuliginosus*	E	LC
Recurvirostridae (Stilts and Avocets)			
Pied Stilt	*Himantopus leucocephalus*	R	LC
Banded Stilt	*Cladorhynchus leucocephalus*	E	LC
Red-necked Avocet	*Recurvirostra novaehollandiae*	E	LC

Common Name	Scientific name	Status	IUCN
Charadriidae (Plovers, Dotterel and Lapwings)			
Grey-headed Lapwing	*Vanellus cinereus*	V	LC
Banded Lapwing	*Vanellus tricolor*	E	LC
Masked Lapwing	*Vanellus miles*	R	LC
Red-kneed Dotterel	*Erythrogonys cinctus*	E	LC
Inland Dotterel	*Peltohyas australis*	E	LC
Pacific Golden Plover	*Pluvialis fulva*	M	LC
American Golden Plover	*Pluvialis dominica*	V	LC
Grey Plover	*Pluvialis squatarola*	M	NT
Common Ringed Plover	*Charadrius hiaticula*	V	LC
Semipalmated Plover	*Charadrius semipalmatus*	V	LC
Little Ringed Plover	*Charadrius dubius*	M	LC
Kentish Plover	*Charadrius alexandrinus*	V	LC
Red-capped Plover	*Charadrius ruficapillus*	R	LC
Double-banded Plover	*Charadrius bicinctus*	M	LC
Lesser Sand Plover	*Charadrius mongolus*	M	EN
Greater Sand Plover	*Charadrius leschenaultii*	M	VU
Caspian Plover	*Charadrius asiaticus*	V	LC
Oriental Plover	*Charadrius veredus*	M	LC
Hooded Plover	*Thinornis cucullatus*	E	VU
Black-fronted Dotterel	*Elseyornis melanops*	R	LC
Rostratulidae (Painted Snipe)			
Australian Painted-snipe	*Rostratula australis*	E	EN
Jacanidae (Jacanas)			
Comb-crested Jacana	*Irediparra gallinacea*	R	LC
Pheasant-tailed Jacana	*Hydrophasianus chirurgus*	V	LC
Pedionomidae (Plains-wanderer)			
Plains-wanderer	*Pedionomus torquatus*	E	CR
Scolopacidae (Snipe, Sandpipers, Stints and allies)			
Latham's Snipe	*Gallinago hardwickii*	M	LC
Pin-tailed Snipe	*Gallinago stenura*	M	LC
Swinhoe's Snipe	*Gallinago megala*	M	LC
Short-billed Dowitcher	*Limnodromus griseus*	V	LC
Long-billed Dowitcher	*Limnodromus scolopaceus*	V	LC
Asian Dowitcher	*Limnodromus semipalmatus*	M	NT
Black-tailed Godwit	*Limosa limosa*	M	NT
Hudsonian Godwit	*Limosa haemastica*	V	LC
Bar-tailed Godwit	*Limosa lapponica*	M	LC
Little Curlew	*Numenius minutus*	M	LC
Whimbrel	*Numenius phaeopus*	M	LC
Eurasian Curlew	*Numenius arquata*	V	LC
Far Eastern Curlew	*Numenius madagascariensis*	M	CR
Upland Sandpiper	*Bartramia longicauda*	V	LC
Terek Sandpiper	*Xenus cinereus*	M	VU
Common Sandpiper	*Actitis hypoleucos*	M	LC
Ruddy Turnstone	*Arenaria interpres*	M	NT
Great Knot	*Calidris tenuirostris*	M	EN
Red Knot	*Calidris canutus*	M	EN
Sanderling	*Calidris alba*	M	LC
Red-necked Stint	*Calidris ruficollis*	M	NT
Little Stint	*Calidris minuta*	V	LC
Long-toed Stint	*Calidris subminuta*	M	LC
Temminck's Stint	*Calidris temminckii*	V	LC
White-rumped Sandpiper	*Calidris fuscicollis*	V	LC
Baird's Sandpiper	*Calidris bairdii*	V	LC
Pectoral Sandpiper	*Calidris melanotos*	M	LC
Sharp-tailed Sandpiper	*Calidris acuminata*	M	LC
Curlew Sandpiper	*Calidris ferruginea*	M	CR

Common Name	Scientific name	Status	IUCN
Dunlin	*Calidris alpina*	V	LC
Stilt Sandpiper	*Calidris himantopus*	V	LC
Broad-billed Sandpiper	*Calidris falcinellus*	M	LC
Buff-breasted Sandpiper	*Calidris subruficollis*	V	LC
Ruff	*Calidris pugnax*	M	LC
Wilson's Phalarope	*Phalaropus tricolor*	V	LC
Red-necked Phalarope	*Phalaropus lobatus*	M	LC
Red Phalarope	*Phalaropus fulicarius*	V	LC
Spotted Redshank	*Tringa erythropus*	V	LC
Common Redshank	*Tringa totanus*	M	LC
Marsh Sandpiper	*Tringa stagnatilis*	M	LC
Lesser Yellowlegs	*Tringa flavipes*	V	LC
Green Sandpiper	*Tringa ochropus*	V	LC
Wood Sandpiper	*Tringa glareola*	M	LC
Grey-tailed Tattler	*Tringa brevipes*	M	LC
Wandering Tattler	*Tringa incana*	M	LC
Common Greenshank	*Tringa nebularia*	M	LC
Nordmann's Greenshank	*Tringa guttifer*	V	LC
Glareolidae (Pratincoles)			
Australian Pratincole	*Stiltia isabella*	R	LC
Oriental Pratincole	*Glareola maldivarum*	M	LC
Laridae (Gulls, Terns and Noddies)			
Brown Noddy	*Anous stolidus*	R	LC
Lesser Noddy	*Anous tenuirostris*	R	EN
Black Noddy	*Anous minutus*	R	LC
Grey Noddy	*Anous albivitta*	R	LC
White Tern	*Gygis alba*	R	LC
Sabine's Gull	*Xema sabini*	V	LC
Silver Gull	*Chroicocephalus novaehollandiae*	R	LC
Black-headed Gull	*Chroicocephalus ridibundus*	V	LC
Laughing Gull	*Leucophaeus atricilla*	V	LC
Franklin's Gull	*Leucophaeus pipixcan*	V	LC
Pacific Gull	*Larus pacificus*	E	LC
Black-tailed Gull	*Larus crassirostris*	V	LC
Mew Gull	*Larus canus*	V	LC
Kelp Gull	*Larus dominicanus*	R	LC
Slaty-backed Gull	*Larus schistisagus*	V	LC
Lesser Black-backed Gull	*Larus fuscus*	V	LC
Gull-billed Tern	*Gelochelidon nilotica*	M	LC
Australian Tern	*Gelochelidon macrotarsa*	BE	LC
Caspian Tern	*Hydroprogne caspia*	R	LC
Greater Crested Tern	*Thalasseus bergii*	R	LC
Lesser Crested Tern	*Thalasseus bengalensis*	R	LC
Little Tern	*Sternula albifrons*	R	LC
Saunders's Tern	*Sternula saundersi*	V	LC
Fairy Tern	*Sternula nereis*	R	VU
Aleutian Tern	*Onychoprion aleuticus*	V	LC
Bridled Tern	*Onychoprion anaethetus*	R	LC
Sooty Tern	*Onychoprion fuscatus*	R	LC
Roseate Tern	*Sterna dougallii*	R	LC
White-fronted Tern	*Sterna striata*	R	NT
Black-naped Tern	*Sterna sumatrana*	R	LC
Common Tern	*Sterna hirundo*	M	LC
Arctic Tern	*Sterna paradisaea*	V	LC
Antarctic Tern	*Sterna vittata*	R	LC
Whiskered Tern	*Chlidonias hybrida*	R	LC
White-winged Tern	*Chlidonias leucopterus*	M	LC
Black Tern	*Chlidonias niger*	V	LC

Common Name	Scientific name	Status	IUCN
Stercorariidae (Skuas and Jaegers)			
South Polar Skua	*Stercorarius maccormicki*	V	LC
Brown Skua	*Stercorarius antarcticus*	R	LC
Pomarine Skua	*Stercorarius pomarinus*	M	LC
Parasitic Jaeger	*Stercorarius parasiticus*	M	LC
Long-tailed Jaeger	*Stercorarius longicaudus*	M	LC
Phaethontidae (Tropicbirds)			
Red-billed Tropicbird	*Phaethon aethereus*	V	LC
Red-tailed Tropicbird	*Phaethon rubricauda*	R	NT
White-tailed Tropicbird	*Phaethon lepturus*	R	EN
Spheniscidae (Penguins)			
King Penguin	*Aptenodytes patagonicus*	R	LC
Emperor Penguin	*Aptenodytes forsteri*	V	LC
Gentoo Penguin	*Pygoscelis papua*	R	NT
Adelic Penguin	*Pygoscelis adeliae*	V	LC
Chinstrap Penguin	*Pygoscelis antarcticus*	V	LC
Fiordland Penguin	*Eudyptes pachyrhynchus*	V	LC
Snares Penguin	*Eudyptes robustus*	V	LC
Erect-crested Penguin	*Eudyptes sclateri*	V	LC
Southern Rockhopper Penguin	*Eudyptes chrysocome*	R	VU
Northern Rockhopper Penguin	*Eudyptes moseleyi*	V	LC
Royal Penguin	*Eudyptes schlegeli*	BE	NT
Macaroni Penguin	*Eudyptes chrysolophus*	R	NT
Little Penguin	*Eudyptula minor*	R	LC
Magellanic Penguin	*Spheniscus magellanicus*	V	LC
Oceanitidae (Southern Storm-Petrels)			
Wilson's Storm Petrel	*Oceanites oceanicus*	R	LC
Grey-backed Storm Petrel	*Garrodia nereis*	R	EN
White-faced Storm Petrel	*Pelagodroma marina*	R	LC
White-bellied Storm Petrel	*Fregetta grallaria*	R	VU
Black-bellied Storm Petrel	*Fregetta tropica*	M	LC
New Zealand Storm Petrel	*Fregetta maoriana*	V	LC
Polynesian Storm Petrel	*Nesofregetta fuliginosa*	V	LC
Diomedeidae (Albatrosses)			
Laysan Albatross	*Phoebastria immutabilis*	V	LC
Wandering Albatross	*Diomedea exulans*	R	CR
Antipodean Albatross	*Diomedea antipodensis*	M	EN
Amsterdam Albatross	*Diomedea amsterdamensis*	V	LC
Tristan Albatross	*Diomedea dabbenena*	V	LC
Southern Royal Albatross	*Diomedea epomophora*	M	VU
Northern Royal Albatross	*Diomedea sanfordi*	M	EN
Sooty Albatross	*Phoebetria fusca*	M	EN
Light-mantled Albatross	*Phoebetria palpebrata*	R	NT
Black-browed Albatross	*Thalassarche melanophris*	R	LC
Campbell Albatross	*Thalassarche impavida*	M	VU
Shy Albatross	*Thalassarche cauta*	BE	VU
Chatham Albatross	*Thalassarche eremita*	V	LC
Salvin's Albatross	*Thalassarche salvini*	M	VU
Grey-headed Albatross	*Thalassarche chrysostoma*	R	EN
Atlantic Yellow-nosed Albatross	*Thalassarche chlororhynchos*	V	LC
Indian Yellow-nosed Albatross	*Thalassarche carteri*	M	EN
Buller's Albatross	*Thalassarche bulleri*	M	NT
Hydrobatidae (Northern Storm-Petrels)			
Swinhoe's Storm-Petrel	*Oceanodroma monorhis*	M	LC
Leach's Storm-Petrel	*Oceanodroma leucorhoa*	V	LC
Band-rumped Storm-Petrel	*Oceanodroma castro*	V	LC
Tristram's Storm-Petrel	*Oceanodroma tristrami*	V	LC
Matsudaira's Storm-Petrel	*Oceanodroma matsudairae*	M	VU

Common Name	Scientific name	Status	IUCN
Procellariidae (Petrels and Shearwaters)			
Southern Giant-Petrel	*Macronectes giganteus*	R	LC
Northern Giant-Petrel	*Macronectes halli*	R	LC
Southern Fulmar	*Fulmarus glacialoides*	M	LC
Antarctic Petrel	*Thalassoica antarctica*	V	LC
Cape Petrel	*Daption capense*	R	LC
Snow Petrel	*Pagodroma nivea*	V	LC
Blue Petrel	*Halobaena caerulea*	R	VU
Broad-billed Prion	*Pachyptila vittata*	M	LC
Salvin's Prion	*Pachyptila salvini*	M	LC
Antarctic Prion	*Pachyptila desolata*	R	LC
Slender-billed Prion	*Pachyptila belcheri*	M	LC
Fairy Prion	*Pachyptila turtur*	R	LC
Fulmar Prion	*Pachyptila crassirostris*	R	LC
Kerguelen Petrel	*Aphrodroma brevirostris*	M	LC
Great-winged Petrel	*Pterodroma macroptera*	BE	LC
Grey-faced Petrel	*Pterodroma gouldi*	M	LC
White-headed Petrel	*Pterodroma lessonii*	R	LC
Atlantic Petrel	*Pterodroma incerta*	V	LC
Providence Petrel	*Pterodroma solandri*	BE	VU
Soft-plumaged Petrel	*Pterodroma mollis*	R	CR
Juan Fernandez Petrel	*Pterodroma externa*	V	LC
Vanuatu Petrel	*Pterodroma occulta*	V	EN
Kermadec Petrel	*Pterodroma neglecta*	R	LC
Herald Petrel	*Pterodroma heraldica*	R	VU
Phoenix Petrel	*Pterodroma alba*	V	EN
Barau's Petrel	*Pterodroma baraui*	V	LC
Mottled Petrel	*Pterodroma inexpectata*	M	NT
White-necked Petrel	*Pterodroma cervicalis*	R	VU
Black-winged Petrel	*Pterodroma nigripennis*	R	LC
Gould's Petrel	*Pterodroma leucoptera*	R	VU
Collared Petrel	*Pterodroma brevipes*	V	LC
Cook's Petrel	*Pterodroma cookii*	V	LC
Stejneger's Petrel	*Pterodroma longirostris*	V	LC
Tahiti Petrel	*Pseudobulweria rostrata*	M	NT
Pycroft's Petrel	*Pterodroma pycrofti*	V	VU
Grey Petrel	*Procellaria cinerea*	R	EN
White-chinned Petrel	*Procellaria aequinoctialis*	M	VU
Black Petrel	*Procellaria parkinsoni*	M	VU
Westland Petrel	*Procellaria westlandica*	M	VU
Streaked Shearwater	*Calonectris leucomelas*	M	NT
Wedge-tailed Shearwater	*Ardenna pacificus*	R	LC
Buller's Shearwater	*Ardenna bulleri*	M	VU
Sooty Shearwater	*Ardenna griseus*	R	NT
Short-tailed Shearwater	*Ardenna tenuirostris*	BE	LC
Pink-footed Shearwater	*Ardenna creatopus*	V	LC
Flesh-footed Shearwater	*Ardenna carneipes*	R	VU
Great Shearwater	*Ardenna gravis*	V	LC
Manx Shearwater	*Puffinus puffinus*	V	LC
Newell's Shearwater	*Puffinus newelli*	V	LC
Fluttering Shearwater	*Puffinus gavia*	M	LC
Hutton's Shearwater	*Puffinus huttoni*	M	EN
Tropical Shearwater	*Puffinus bailloni*	V	LC
Heinroth's Shearwater	*Puffinus heinrothi*	V	LC
Little Shearwater	*Puffinus assimilis*	R	LC
Subantarctic Sheawater	*Puffinus elegans*	M	LC
South Georgian Diving-Petrel	*Pelecanoides georgicus*	R	LC
Common Diving-Petrel	*Pelecanoides urinatrix*	R	LC

Common Name	Scientific name	Status	IUCN
Bulwer's Petrel	*Bulweria bulwerii*	M	LC
Jouanin's Petrel	*Bulweria fallax*	V	LC
Ciconiidae (Storks)			
Black-necked Stork	*Ephippiorhynchus asiaticus*	R	LC
Fregatidae (Frigatebirds)			
Christmas Frigatebird	*Fregata andrewsi*	BE	CR
Great Frigatebird	*Fregata minor*	R	LC
Lesser Frigatebird	*Fregata ariel*	R	LC
Sulidae (Gannets and Boobies)			
Cape Gannet	*Morus capensis*	V	LC
Australasian Gannet	*Morus serrator*	R	LC
Abbott's Booby	*Papasula abbotti*	BE	EN
Masked Booby	*Sula dactylatra*	R	LC
Red-footed Booby	*Sula sula*	R	LC
Brown Booby	*Sula leucogaster*	R	LC
Phalacrocoracidae (Cormorants and Shags)			
Little Pied Cormorant	*Microcarbo melanoleucos*	R	LC
Black-faced Cormorant	*Phalacrocorax fuscescens*	E	LC
Little Black Cormorant	*Phalacrocorax sulcirostris*	R	LC
Australian Pied Cormorant	*Phalacrocorax varius*	R	LC
Great Cormorant	*Phalacrocorax carbo*	R	LC
Heard Island Shag	*Leucocarbo nivalis*	E	LC
Macquarie Shag	*Leucocarbo purpurascens*	E	LC
Anhingidae (Darter)			
Australasian Darter	*Anhinga novaehollandiae*	R	LC
Threskiornithidae (Ibis and Spoonbills)			
Australian White Ibis	*Threskiornis moluccus*	R	LC
Straw-necked Ibis	*Threskiornis spinicollis*	BE	LC
Glossy Ibis	*Plegadis falcinellus*	R	LC
Royal Spoonbill	*Platalea regia*	R	LC
Yellow-billed Spoonbill	*Platalea flavipes*	E	LC
Ardeidae (Herons, Egrets and Bitterns)			
Australasian Bittern	*Botaurus poiciloptilus*	R	EN
Black-backed Bittern	*Ixobrychus dubius*	R	LC
Yellow Bittern	*Ixobrychus sinensis*	V	LC
Von Schrenck's Bittern	*Ixobrychus eurhythmus*	V	LC
Cinnamon Bittern	*Ixobrychus cinnamomeus*	V	LC
Black Bittern	*Dupetor flavicollis*	R	LC
Japanese Night-Heron	*Gorsachius goisagi*	V	LC
Malayan Night-Heron	*Gorsachius melanolophus*	V	LC
Black-crowned Night-Heron	*Nycticorax nycticorax*	V	LC
Nankeen Night-Heron	*Nycticorax caledonicus*	R	LC
Striated Heron	*Butorides striata*	R	LC
Chinese Pond Heron	*Ardeola bacchus*	V	LC
Javan Pond Heron	*Ardeola speciosa*	V	LC
Eastern Cattle Egret	*Bubulcus coromandus*	R	LC
Grey Heron	*Ardea cinerea*	V	LC
White-necked Heron	*Ardea pacifica*	E	LC
Great-billed Heron	*Ardea sumatrana*	R	LC
Purple Heron	*Ardea purpurea*	V	LC
Great Egret	*Ardea alba*	R	LC
Intermediate Egret	*Egretta intermedia*	R	LC
Pied Heron	*Egretta picata*	R	LC
White-faced Heron	*Egretta novaehollandiae*	R	LC
Little Egret	*Egretta garzetta*	R	LC
Western Reef Heron	*Egretta gularis*	V	LC
Pacific Reef Heron	*Egretta sacra*	R	LC
Pelicanidae (Pelican)			

Common Name	Scientific name	Status	IUCN
Australian Pelican	*Pelecanus conspicillatus*	BE	LC
Pandionidae (Ospreys)			
Western Osprey	*Pandion haliaetus*	V	LC
Eastern Osprey	*Pandion cristatus*	R	LC
Accipitridae (Kites, Hawks and Eagles)			
Black-shouldered Kite	*Elanus axillaris*	E	LC
Letter-winged Kite	*Elanus scriptus*	E	NT
Oriental Honey-buzzard	*Pernis ptilorhynchus*	V	LC
Square-tailed Kite	*Lophoictinia isura*	E	LC
Black-breasted Buzzard	*Hamirostra melanosternon*	E	LC
Pacific Baza	*Aviceda subcristata*	R	LC
Little Eagle	*Hieraaetus morphnoides*	E	LC
Gurney's Eagle	*Aquila gurneyi*	V	LC
Wedge-tailed Eagle	*Aquila audax*	R	LC
Red Goshawk	*Erythrotriorchis radiatus*	E	NT
Chinese Sparrowhawk	*Accipiter soloensis*	V	LC
Grey Goshawk	*Accipiter novaehollandiae*	E	LC
Brown Goshawk	*Accipiter fasciatus*	R	LC
Japanese Sparrowhawk	*Accipiter gularis*	V	LC
Collared Sparrowhawk	*Accipiter cirrocephalus*	R	LC
Swamp Harrier	*Circus approximans*	R	LC
Spotted Harrier	*Circus assimilis*	E	LC
Black Kite	*Milvus migrans*	R	LC
Whistling Kite	*Haliastur sphenurus*	R	LC
Brahminy Kite	*Haliastur indus*	R	LC
White-bellied Sea-Eagle	*Haliaeetus leucogaster*	R	LC
Tytonidae (Barn Owls)			
Greater Sooty Owl	*Tyto tenebricosa*	R	LC
Lesser Sooty Owl	*Tyto multipunctata*	E	LC
Masked Owl	*Tyto novaehollandiae*	R	LC
Barn Owl	*Tyto javanica*	R	LC
Eastern Grass Owl	*Tyto longimembris*	R	LC
Strigidae (Owls)			
Oriental Scops-owl	*Otus sunia*	V	LC
Buffy Fish-Owl	*Ketupa ketupu*	V	LC
Rufous Owl	*Ninox rufa*	R	LC
Powerful Owl	*Ninox strenua*	E	LC
Barking Owl	*Ninox connivens*	R	LC
Southern Boobook	*Ninox boobook*	R	LC
Morepork	*Ninox novaeseelandiae*	R	CR
Northern Boobook	*Ninox japonica*	V	LC
Christmas Island Boobook	*Ninox natalis*	E	VU
Upupidae (Hoopoes)			
Eurasian Hoopoe	*Upupa epops*	V	LC
Coraciidae (Dollarbirds)			
European Roller	*Coracias garrulus*	V	LC
Oriental Dollarbird	*Eurystomus orientalis*	R	LC
Alcedinidae (Kingfishers)			
Little Paradise-Kingfisher	*Tanysiptera hydrocharis*	V	LC
Buff-breasted Paradise-Kingfisher	*Tanysiptera sylvia*	R	LC
Laughing Kookaburra	*Dacelo novaeguineae*	E	LC
Blue-winged Kookaburra	*Dacelo leachii*	R	LC
Black-capped Kingfisher	*Halcyon pileata*	V	LC
Forest Kingfisher	*Todiramphus macleayii*	R	LC
Torresian Kingfisher	*Todiramphus sordidus*	R	LC
Sacred Kingfisher	*Todiramphus sanctus*	R	LC
Collared Kingfisher	*Todiramphus chloris*	R	LC
Red-backed Kingfisher	*Todiramphus pyrrhopygius*	E	LC

Common Name	Scientific name	Status	IUCN
Yellow-billed Kingfisher	Syma torotoro	R	LC
Common Kingfisher	Alcedo atthis	V	LC
Azure Kingfisher	Ceyx azureus	R	LC
Little Kingfisher	Ceyx pusillus	R	LC
Meropidae (Bee-eaters)			
Rainbow Bee-eater	Merops ornatus	R	LC
Falconidae (Falcons)			
Nankeen Kestrel	Falco cenchroides	R	LC
Eurasian Hobby	Falco subbuteo	V	LC
Australian Hobby	Falco longipennis	R	LC
Brown Falcon	Falco berigora	R	LC
Grey Falcon	Falco hypoleucos	BE	VU
Black Falcon	Falco subniger	E	LC
Peregrine Falcon	Falco peregrinus	R	LC
Cacatuidae (Cockatoos)			
Palm Cockatoo	Probosciger aterrimus	R	VU
Red-tailed Black-Cockatoo	Calyptorhynchus banksii	E	LC
Glossy Black-Cockatoo	Calyptorhynchus lathami	E	LC
Yellow-tailed Black-Cockatoo	Calyptorhynchus funereus	E	LC
Carnaby's Black-Cockatoo	Calyptorhynchus latirostris	E	EN
Baudin's Black-Cockatoo	Calyptorhynchus baudinii	E	EN
Gang-gang Cockatoo	Callocephalon fimbriatum	E	LC
Major Mitchell's Cockatoo	Lophochroa leadbeateri	E	LC
Galah	Eolophus roseicapilla	E	LC
Long-billed Corella	Cacatua tenuirostris	E	LC
Western Corella	Cacatua pastinator	E	LC
Little Corella	Cacatua sanguinea	R	LC
Sulphur-crested Cockatoo	Cacatua galerita	R	LC
Cockatiel	Nymphicus hollandicus	E	LC
Psittaculidae (Parrots)			
Superb Parrot	Polytelis swainsonii	E	LC
Regent Parrot	Polytelis anthopeplus	E	LC
Princess Parrot	Polytelis alexandrae	E	NT
Australian King-Parrot	Alisterus scapularis	E	LC
Red-winged Parrot	Aprosmictus erythropterus	R	LC
Eclectus Parrot	Eclectus roratus	R	NT
Red-cheeked Parrot	Geoffroyus geoffroyi	R	LC
Red-rumped Parrot	Psephotus haematonotus	E	LC
Eastern Bluebonnet	Northiella haematogaster	E	LC
Naretha Bluebonnet	Northiella narethae	E	LC
Mulga Parrot	Psephotus varius	E	LC
Hooded Parrot	Psephotus dissimilis	E	LC
Golden-shouldered Parrot	Psephotus chrysopterygius	E	EN
Red-capped Parrot	Purpureicephalus spurius	E	LC
Green Rosella	Platycercus caledonicus	E	LC
Crimson Rosella	Platycercus elegans	E	LC
Northern Rosella	Platycercus venustus	E	LC
Pale-headed Rosella	Platycercus adscitus	E	LC
Eastern Rosella	Platycercus eximius	E	LC
Western Rosella	Platycercus icterotis	E	LC
Australian Ringneck	Barnardius zonarius	E	LC
Swift Parrot	Lathamus discolor	E	CR
Norfolk Parakeet	Cyanoramphus cookii	R	CR
Eastern Ground Parrot	Pezoporus wallicus	E	LC
Western Ground Parrot	Pezoporus flaviventris	E	CR
Night Parrot	Pezoporus occidentalis	E	EN
Bourke's Parrot	Neopsephotus bourkii	E	LC
Blue-winged Parrot	Neophema chrysostoma	E	LC

Common Name	Scientific name	Status	IUCN
Elegant Parrot	*Neophema elegans*	E	LC
Rock Parrot	*Neophema petrophila*	E	LC
Orange-bellied Parrot	*Neophema chrysogaster*	E	CR
Turquoise Parrot	*Neophema pulchella*	E	LC
Scarlet-chested Parrot	*Neophema splendida*	E	LC
Little Lorikeet	*Glossopsitta pusilla*	E	LC
Purple-crowned Lorikeet	*Glossopsitta porphyrocephala*	E	LC
Varied Lorikeet	*Psitteuteles versicolor*	E	LC
Coconut Lorikeet	*Trichoglossus haematodus*	M	LC
Rainbow Lorikeet	*Trichoglossus moluccanus*	E	LC
Red-collared Lorikeet	*Trichoglossus rubritorquis*	E	LC
Scaly-breasted Lorikeet	*Trichoglossus chlorolepidotus*	E	LC
Musk Lorikeet	*Glossopsitta concinna*	E	LC
Budgerigar	*Melopsittacus undulatus*	E	LC
Double-eyed Fig Parrot	*Cyclopsitta diophthalma*	R	LC
Pittidae (Pittas)			
Papuan Pitta	*Erythropitta macklotii*	R	LC
Hooded Pitta	*Pitta sordida*	V	LC
Fairy Pitta	*Pitta nympha*	V	LC
Blue-winged Pitta	*Pitta moluccensis*	V	LC
Rainbow Pitta	*Pitta iris*	E	LC
Noisy Pitta	*Pitta versicolor*	R	LC
Menuridae (Lyrebirds)			
Albert's Lyrebird	*Menura alberti*	E	NT
Superb Lyrebird	*Menura novaehollandiae*	E	LC
Atrichornithidae (Scrub-birds)			
Rufous Scrub-bird	*Atrichornis rufescens*	E	EN
Noisy Scrub-bird	*Atrichornis clamosus*	E	EN
Ptilonorhynchidae (Bowerbirds and Catbirds)			
Green Catbird	*Ailuroedus crassirostris*	E	LC
Spotted Catbird	*Ailuroedus maculosus*	E	LC
Black-eared Catbird	*Ailuroedus melanotis*	R	LC
Tooth-billed Bowerbird	*Scenopoeetes dentirostris*	E	LC
Golden Bowerbird	*Prionodura newtoniana*	E	LC
Regent Bowerbird	*Sericulus chrysocephalus*	E	LC
Satin Bowerbird	*Ptilonorhynchus violaceus*	E	LC
Western Bowerbird	*Chlamydera guttata*	E	LC
Great Bowerbird	*Chlamydera nuchalis*	E	LC
Spotted Bowerbird	*Chlamydera maculata*	E	LC
Fawn-breasted Bowerbird	*Chlamydera cerviniventris*	R	NT
Climacteridae (Australasian Treecreepers)			
White-throated Treecreeper	*Cormobates leucophaea*	E	LC
Red-browed Treecreeper	*Climacteris erythrops*	E	LC
White-browed Treecreeper	*Climacteris affinis*	E	LC
Rufous Treecreeper	*Climacteris rufus*	E	LC
Brown Treecreeper	*Climacteris picumnus*	E	LC
Black-tailed Treecreeper	*Climacteris melanurus*	E	LC
Maluridae (Australasian Wrens)			
Lovely Fairy-wren	*Malurus amabilis*	E	LC
Variegated Fairy-wren	*Malurus lamberti*	E	LC
Purple-backed Fairy-wren	*Malurus assimilis*	E	LC
Blue-breasted Fairy-wren	*Malurus pulcherrimus*	E	LC
Red-winged Fairy-wren	*Malurus elegans*	E	LC
Superb Fairy-wren	*Malurus cyaneus*	E	LC
Splendid Fairy-wren	*Malurus splendens*	E	LC
Purple-crowned Fairy-wren	*Malurus coronatus*	E	LC
Red-backed Fairy-wren	*Malurus melanocephalus*	E	LC
White-winged Fairy-wren	*Malurus leucopterus*	E	LC

Common Name	Scientific name	Status	IUCN
Southern Emu-wren	*Stipiturus malachurus*	E	LC
Mallee Emu-wren	*Stipiturus mallee*	E	EN
Rufous-crowned Emu-wren	*Stipiturus ruficeps*	E	LC
Grey Grasswren	*Amytornis barbatus*	E	LC
Black Grasswren	*Amytornis housei*	E	NT
White-throated Grasswren	*Amytornis woodwardi*	E	VU
Carpentarian Grasswren	*Amytornis dorotheae*	E	VU
Short-tailed Grasswren	*Amytornis merrotsyi*	E	VU
Striated Grasswren	*Amytornis striatus*	E	LC
Eyrean Grasswren	*Amytornis goyderi*	E	LC
Western Grasswren	*Amytornis textilis*	E	LC
Thick-billed Grasswren	*Amytornis modestus*	E	LC
Dusky Grasswren	*Amytornis purnelli*	E	LC
Kalkadoon Grasswren	*Amytornis ballarae*	E	LC
Meliphagidae (Honeyeaters)			
Black Honeyeater	*Sugomel nigrum*	E	LC
Dusky Honeyeater	*Myzomela obscura*	R	LC
Red-headed Honeyeater	*Myzomela erythrocephala*	R	LC
Scarlet Honeyeater	*Myzomela sanguinolenta*	R	LC
Tawny-crowned Honeyeater	*Gliciphila melanops*	E	LC
Green-backed Honeyeater	*Glycichaera fallax*	R	LC
Eastern Spinebill	*Acanthorhynchus tenuirostris*	E	LC
Western Spinebill	*Acanthorhynchus superciliosus*	E	LC
Pied Honeyeater	*Certhionyx variegatus*	E	LC
Banded Honeyeater	*Cissomela pectoralis*	E	LC
Brown Honeyeater	*Lichmera indistincta*	R	LC
Crescent Honeyeater	*Phylidonyris pyrrhopterus*	E	LC
New Holland Honeyeater	*Phylidonyris novaehollandiae*	E	LC
White-cheeked Honeyeater	*Phylidonyris niger*	E	LC
White-streaked Honeyeater	*Trichodere cockerelli*	E	LC
Painted Honeyeater	*Grantiella picta*	E	VU
Striped Honeyeater	*Plectorhyncha lanceolata*	E	LC
Macleay's Honeyeater	*Xanthotis macleayanus*	E	LC
Tawny-breasted Honeyeater	*Xanthotis flaviventer*	R	LC
Little Friarbird	*Philemon citreogularis*	R	LC
Helmeted Friarbird	*Philemon buceroides*	R	LC
Hornbill Friarbird	*Philemon yorki*	E	LC
Silver-crowned Friarbird	*Philemon argenticeps*	E	LC
Noisy Friarbird	*Philemon corniculatus*	BE	LC
Blue-faced Honeyeater	*Entomyzon cyanotis*	R	LC
Black-chinned Honeyeater	*Melithreptus gularis*	E	NT
Strong-billed Honeyeater	*Melithreptus validirostris*	E	LC
Brown-headed Honeyeater	*Melithreptus brevirostris*	E	LC
White-throated Honeyeater	*Melithreptus albogularis*	R	LC
White-naped Honeyeater	*Melithreptus lunatus*	E	LC
Gilbert's Honeyeater	*Melithreptus chloropsis*	E	LC
Black-headed Honeyeater	*Melithreptus affinis*	E	LC
White-eared Honeyeater	*Nesoptilotis leucotis*	E	LC
Yellow-throated Honeyeater	*Nesoptilotis flavicollis*	E	LC
Gibberbird	*Ashbyia lovensis*	E	LC
Crimson Chat	*Epthianura tricolor*	E	LC
Orange Chat	*Epthianura aurifrons*	E	LC
Yellow Chat	*Epthianura crocea*	E	LC
White-fronted Chat	*Epthianura albifrons*	E	LC
Rufous-banded Honeyeater	*Conopophila albogularis*	R	LC
Rufous-throated Honeyeater	*Conopophila rufogularis*	E	LC
Grey Honeyeater	*Conopophila whitei*	E	LC
Brown-backed Honeyeater	*Ramsayornis modestus*	R	LC

Common Name	Scientific name	Status	IUCN
Bar-breasted Honeyeater	Ramsayornis fasciatus	E	LC
Spiny-cheeked Honeyeater	Acanthagenys rufogularis	E	LC
Little Wattlebird	Anthochaera chrysoptera	E	LC
Western Wattlebird	Anthochaera lunulata	E	LC
Red Wattlebird	Anthochaera carunculata	E	LC
Yellow Wattlebird	Anthochaera paradoxa	E	LC
Regent Honeyeater	Anthochaera phrygia	E	CR
Bridled Honeyeater	Bolemoreus frenatus	E	LC
Eungella Honeyeater	Bolemoreus hindwoodi	E	NT
Yellow-faced Honeyeater	Caligavis chrysops	E	LC
Yellow-tufted Honeyeater	Lichenostomus melanops	E	LC
Purple-gaped Honeyeater	Lichenostomus cratitius	E	LC
Bell Miner	Manorina melanophrys	E	LC
Noisy Miner	Manorina melanocephala	E	LC
Yellow-throated Miner	Manorina flavigula	E	LC
Black-eared Miner	Manorina melanotis	E	EN
White-fronted Honeyeater	Purnella albifrons	E	LC
White-gaped Honeyeater	Stomiopera unicolor	E	LC
Yellow Honeyeater	Stomiopera flava	E	LC
Varied Honeyeater	Gavicalis versicolor	R	LC
Mangrove Honeyeater	Gavicalis fasciogularis	E	LC
Singing Honeyeater	Gavicalis virescens	E	LC
Yellow-tinted Honeyeater	Ptilotula flavescens	E	LC
Fuscous Honeyeater	Ptilotula fusca	E	LC
Grey-headed Honeyeater	Ptilotula keartlandi	E	LC
Grey-fronted Honeyeater	Ptilotula plumula	E	LC
Yellow-plumed Honeyeater	Ptilotula ornata	E	LC
White-plumed Honeyeater	Ptilotula penicillata	E	LC
Graceful Honeyeater	Meliphaga gracilis	R	LC
White-lined Honeyeater	Meliphaga albilineata	E	LC
Kimberley Honeyeater	Meliphaga fordiana	E	LC
Yellow-spotted Honeyeater	Meliphaga notata	E	LC
Lewin's Honeyeater	Meliphaga lewinii	E	LC
Dasyornithidae (Bristlebirds)			
Eastern Bristlebird	Dasyornis brachypterus	E	EN
Western Bristlebird	Dasyornis longirostris	E	EN
Rufous Bristlebird	Dasyornis broadbenti	E	LC
Pardalotidae (Pardalotes)			
Spotted Pardalote	Pardalotus punctatus	E	LC
Forty-spotted Pardalote	Pardalotus quadragintus	E	EN
Red-browed Pardalote	Pardalotus rubricatus	E	LC
Striated Pardalote	Pardalotus striatus	E	LC
Acanthizidae (Australasian Warblers)			
Pilotbird	Pycnoptilus floccosus	E	LC
Scrubtit	Acanthornis magna	E	LC
Rockwarbler	Origma solitaria	E	LC
Chestnut-rumped Heathwren	Hylacola pyrrhopygia	E	LC
Shy Heathwren	Hylacola cauta	E	LC
Striated Fieldwren	Calamanthus fuliginosus	E	LC
Western Fieldwren	Calamanthus montanellus	E	LC
Rufous Fieldwren	Calamanthus campestris	E	LC
Redthroat	Pyrrholaemus brunneus	E	LC
Speckled Warbler	Pyrrholaemus sagittatus	E	LC
Fernwren	Oreoscopus gutturalis	E	LC
Atherton Scrubwren	Sericornis keri	E	LC
White-browed Scrubwren	Sericornis frontalis	E	LC
Spotted Scrubwren	Sericornis maculatus	E	LC
Tasmanian Scrubwren	Sericornis humilis	E	LC

Common Name	Scientific name	Status	IUCN
Yellow-throated Scrubwren	Sericornis citreogularis	E	LC
Large-billed Scrubwren	Sericornis magnirostra	E	LC
Tropical Scrubwren	Sericornis beccarii	R	LC
Weebill	Smicrornis brevirostris	E	LC
Brown Gerygone	Gerygone mouki	E	LC
Norfolk Gerygone	Gerygone modesta	E	NT
Mangrove Gerygone	Gerygone levigaster	R	LC
Western Gerygone	Gerygone fusca	E	LC
Dusky Gerygone	Gerygone tenebrosa	E	LC
Large-billed Gerygone	Gerygone magnirostris	R	LC
Green-backed Gerygone	Gerygone chloronota	R	LC
White-throated Gerygone	Gerygone olivacea	R	LC
Fairy Gerygone	Gerygone palpebrosa	R	LC
Mountain Thornbill	Acanthiza katherina	E	LC
Brown Thornbill	Acanthiza pusilla	E	LC
Inland Thornbill	Acanthiza apicalis	E	LC
Tasmanian Thornbill	Acanthiza ewingii	E	LC
Chestnut-rumped Thornbill	Acanthiza uropygialis	E	LC
Buff-rumped Thornbill	Acanthiza reguloides	E	LC
Western Thornbill	Acanthiza inornata	E	LC
Slender-billed Thornbill	Acanthiza iredalei	E	LC
Yellow-rumped Thornbill	Acanthiza chrysorrhoa	E	LC
Yellow Thornbill	Acanthiza nana	E	LC
Striated Thornbill	Acanthiza lineata	E	LC
Slaty-backed Thornbill	Acanthiza robustirostris	E	LC
Southern Whiteface	Aphelocephala leucopsis	E	LC
Chestnut-breasted Whiteface	Aphelocephala pectoralis	E	NT
Banded Whiteface	Aphelocephala nigricincta	E	LC
Pomatostomidae (Australasian Babblers)			
Grey-crowned Babbler	Pomatostomus temporalis	R	LC
Hall's Babbler	Pomatostomus halli	E	LC
White-browed Babbler	Pomatostomus superciliosus	E	LC
Chestnut-crowned Babbler	Pomatostomus ruficeps	E	LC
Orthonychidae (Logrunners)			
Australian Logrunner	Orthonyx temminckii	E	LC
Chowchilla	Orthonyx spaldingii	E	LC
Psophodidae (Whipbirds)			
Eastern Whipbird	Psophodes olivaceus	E	LC
Mallee Western Whipbird	Psophodes leucogaster	E	VU
Heath Western Whipbird	Psophodes nigrogularis	E	EN
Chirruping Wedgebill	Psophodes cristatus	E	LC
Chiming Wedgebill	Psophodes occidentalis	E	LC
Cinclosomatidae (Quail-thrushes)			
Spotted Quail-thrush	Cinclosoma punctatum	E	LC
Chestnut Quail-thrush	Cinclosoma castanotum	E	LC
Copperback Quail-thrush	Cinclosoma clarum	E	LC
Cinnamon Quail-thrush	Cinclosoma cinnamomeum	E	LC
Nullarbor Quail-thrush	Cinclosoma alisteri	E	LC
Chestnut-breasted Quail-thrush	Cinclosoma castaneothorax	E	LC
Western Quail-thrush	Cinclosoma marginatum	E	LC
Machaerirhynchidae (Boatbills)			
Yellow-breasted Boatbill	Machaerirhynchus flaviventer	R	LC
Artamidae (Woodswallows, Butcherbirds and Allies)			
White-breasted Woodswallow	Artamus leucorynchus	R	LC
Masked Woodswallow	Artamus personatus	E	LC
White-browed Woodswallow	Artamus superciliosus	E	LC
Black-faced Woodswallow	Artamus cinereus	E	LC
Dusky Woodswallow	Artamus cyanopterus	E	LC